THE SUBTERRANEANS

THE SUBTERRANEANS

by
M. D. Willis

KROSHKA BOOKS
COMMACK, NY

Creative Design: Gavin Aghamore
Editorial Production: Susan Boriotti
Art Director: Maria Ester Hawrys
Assistant Director: Elenor Kallberg
Graphics: Frank Grucci
Manuscript Coordinator: Phyllis Gaynor
Book Production: Joanne Bennette, Michele Keller
Christine Mathosian and Tammy Sauter
Circulation: Cathy DeGregory and Annette Hellinger

Library of Congress Cataloging-in-Publication Data available upon request

ISBN 1-56072-339-4

Kroshka Books. A division of
Nova Science Publishers, Inc.
6080 Jericho Turnpike, Suite 207
Commack, New York 11725
Tele. 516-499-3103 Fax 516-499-3146
E Mail Novascience@earthlink.net

Printed in the United States of America

CONTENTS

PREFACE

The idea that there may be humanoid creatures existing in some unknown locations beneath the surface of the Earth is so fantastic that it seems absurd. How could that possibly be? After all, the ground is pretty much solid mass and what we know of it gives us no reason to suppose that any such thing is remotely possible. When this notion occurred to me many years ago, this was my rational reaction. The old science fiction image of an Earth that would allow journeys deep into its center is patently impossible in light of contemporary scientific understanding. I gave it no further thought. Over the intervening years, with the revolution that has taken place in the understanding of Earth's structure, with the discovery of new and ever more unusual forms of life, and startling new Archaeological, Oceanographic and Geological discoveries, it seemed that stranger things have been found. Perhaps the idea seemed so fantastic because earlier scientific models of the Earth's history and structure were much cruder and simpler. Perhaps the underground is far more varied and complex than once thought. Perhaps the idea was not so absurd after all.

That all of the continents were once a part of a gigantic supercontinent, Pangea, that ruptured along great jagged lines and drifted far apart to their present locations is one of the most fantastic ideas imaginable, one that seemed like pure science fiction at the time it was introduced by Alfred Wegener in 1912, yet is it so. And throughout these massive worldwide changes, some form of life has adapted itself to survive virtually every inch of our planet. Creatures inhabit or have inhabited all the lands, the air, the seas at all depths and extremes of temperature and pressure and in the soil itself. Other forms of animal life have become isolated from the surface and developed unique adaptations to underground environments, why not humans? Life itself is fantastic, and most fantastic of all is that, in some form or other, it has managed to survive practically everywhere. I began to reexamine the thought; to challenge its validity in light of current scientific knowledge and develop hypothetical scenarios that might be possible.

Of course believing, or even proving that a thing is possible does not make it a fact. Only in time, and with greater knowledge of the Earth and of human origins and development can we know. It may be that the ideas presented in this book will prove wrong, or that only some small aspect is correct. I urge you to keep an open mind, however. These speculations should be kept a part of our thinking until they are established one way or the other. All plausible scenarios about the development and dissemination of life on Earth should be entertained until such time as they are demonstrated inaccurate or impossible. Doing so will stimulate our curiosity and enrich the intellectual environment so that we may better understand our planet and its people.

THE FLUID DEEP

In 1960, when the Bathyscaphe Trieste descended to the depths of the Marianas trench in the western Pacific Ocean, the deepest known point on Earth, something quite unexpected was found. At the bottom of the abyss, some seven miles below sea level, at the crushing pressure of 1,100 atmospheres, extraordinary forms of life were discovered. These were not merely simple single-celled plants or animals, the most that might be expected under such conditions, but sophisticated higher organisms - sizable swimming fish unlike any found nearer the surface. They were, for the most part, recognizable as relatives of known species; relatives that had mutated to adapt to a life at extreme pressures and in complete darkness. A few were found to be related to species thought to be extinct, and identification made only through similarities to fossil remains. So adapted to the extreme conditions were these forms of life that they could not exist in waters near the surface. Their evolutionary history was completely isolated from other life on our planet, even from their own progenitors.

In the mid 1800's it was generally held that the greatest depths at which life might be found was about 1300-1400 feet. This based largely on the fact that sunlight does not penetrate any further than that and so plant life would not be possible, and without plant life, animal life could not exist. Within ten years however, in 1860, a starfish was dredged up from a depth of 7500 feet, disproving that assumption.

The beginnings of modern oceanography only began in the 1870's with the four-year long voyage of the British H.M.S. Challenger which recorded depth soundings over an 80,000 mile course that took it throughout all the world's oceans. It was this research that first indicated the presence of extensive undersea ridges, but which were not actually to be seen by humans until 1973, nearly a hundred years later, and only after man had first set foot on the moon.

In the 17th century, the Earth was thought to be a mere 6,000 years old. One biblical scholar even placed the exact time and date of creation at 9:00 a.m., October 26, 4004 BC Although this sounds ridiculous to us now, it was a consensus view for almost 200 years. It was not until 1795 that James Hutton, the father of modern geology, proposed that the Earth was much older - hundreds of thousands of years old. The rock stratifications he observed that supported his assertions are now known to have taken more on the order of 70 million years to form and our best modern estimates now place the Earth's age at around 4.6 billion years.

The history of our understanding of the physiology of Earth and the remarkable diversity of life which populates it has been filled with ever-expanding discoveries and the repeated revision or abandonment of previously held models. The Earth is a continually-evolving complex system which is still undergoing constant change and living organisms are an intimate part of this process, both being modified by the environment and changing the environment as well. The nature and evolution of living things cannot be separated from the environment and the physical actions which shape the planet.

The process of Continental Drift (a theory of continental evolution that is now accepted as fact but as recently as 1960 was considered radical and unlikely) forms deep gashes, crevices deeper than any such formations on the surface of the Earth, in the oceans' floors as Tectonic Plates are forced under one another, disappearing forever into the interior of the Earth.

Deeper both in terms of the variation from the bottommost and topmost boundaries of the formation and also, since the ocean trenches occur well below the surface of the oceans, are much, much deeper in absolute terms of distance from the Earth's surface. The Marianas trench is not the only such formation. There are others: notably the Tonga-Kermadec trench north of New Zealand, the Peru-Chile trench along the western coast of South America, the Puerto Rico trench in the western Atlantic, as well as the

Japan, Java, Mindanao and Aleutian trenches. There are, in all, 17 deep trenches ranging in depth from 24,442 feet (the Sunda trench) to 35,600 feet (the Marianas trench) in the Atlantic, Pacific and Indian Oceans.

It is in these locations, under pressures as high as 3 1/2 tons per square inch, in absolute darkness, at temperatures ranging from 28 to 38 degrees Fahrenheit (salt water freezes at lower temperatures than fresh water), and with much less Oxygen than in waters nearer the surface, that the some of the strangest forms of life yet found on our planet exist. A great many resemble aircraft flying at night more than fishes swimming in water, for the majority carry their own running lights in the form of bioluminescent patches over portions of their bodies. The actual skin and tissue of many is translucent or completely transparent with the inner organs visible, and appear much like an fluoroscopic image. Many are blind, with eyes as mere vestigial reminders of their links to surface relatives. Others have vision so poor as to be able to see only gross shapes and patches of light. The glowing lights carried by many of these fish are lures to entice prey closer. Some lights are actually dangled from a fishing pole-like appendage just over the predator's mouth, or in some cases are actually inside to mouth to entice unsuspecting prey.

There are also examples of life in the sea (at much shallower depths) which show virtually no recent evolution. In 1938, a specimen of a Coelacanth fish was caught by local fishermen off the coast of South Africa. This fish, very much alive, was thought to have been extinct for 70 million years. More living specimens of this fish have since been caught in the waters around South Africa.

Other plants and animals, well known because of their fossil remains found in various places throughout the world, have occasionally turned up in various oceans. In 1952, off the Mexican Pacific coast, several samples of an ancient mollusk (half clam and half snail) were recovered in just under 12,000 of water - alive. It too was thought to have vanished from the Earth 350 million years ago.

Exactly how and why some living creatures manage to escape the evolutionary fate of others of their kind and remain unchanged for so long is unknown. It appears though, that in some areas of the sea they become separated from the mainstream of their species and reproduce without being biologically influenced by the rest. Perhaps in the areas in which they become isolated, they are well enough suited to the prevailing conditions, or those conditions are less competitive, so that significant change is not necessary for survival.

We think of water as a relatively consistent and uniform medium with little variation in its composition. However as a natural habitat, water is far from uniform. Variations in temperature, salinity, pressure, pH, light, mineral content and oxygen content constitute significant factors which influence the evolution of life forms. Natural selection responds to the subtlest differences in environment, and though discernible change in a species takes tremendously long, results in substantial differences among organisms in what, superficially, may seem like the same habitat. Even so seemingly uniform a medium as water can present, from a biological point of view, great enough variations of environment (so much so as to produce actual isolation) to produce great biological diversity.

The oceans have long been considered the incubators of the first life on Earth. Exactly how this could have happened has only recently been demonstrated experimentally. Under laboratory conditions which duplicated the environment thought to have existed three to four billion years ago, simple amino acids, the foundations of life, spontaneously formed out of the interactions of ultraviolet radiation, electric sparks, Nitrogen and Carbon Dioxide. Furthermore, these coalesced into cells which eventually developed the ability to reproduce themselves, just as the first spontaneously-occurring life must have. As these cells developed the mechanism of photosynthesis to extract energy from sunlight, they gave off Oxygen. The gradual accumulation of free Oxygen in the atmosphere, where none had existed before, transformed the Earth's environment into one which allowed the development of air-breathing animal life. Thus it was the formation of life which created the atmosphere we now have.

Paradoxically, it is the oceans, which provided conditions out of which life first grew, that also hold some of the oddest forms of life known to man - a result of adaptations to the great depths which are some of the most hostile environments on Earth. And even stranger animals than those already described have been found. As we shall see in later chapters, ocean ridges hold entirely new forms of life including bacteria which can live and reproduce at temperatures well above the boiling point of water and clams which can reproduce 500 times faster than any others known.

And these great varieties of creatures are possible just within the watery environments. When we expand the possibilities of life adaptations to all earthly environments, even greater variation results: Nature abounds with examples of animals (and plants) adapting to extremes of conditions. That life adapts through evolution to a wide variety of environments is such an accepted fact that we tend to take it for granted. We're accustomed to seeing birds in the air and fish (and even mammals) in the sea. But what could be more fantastic than for reptiles to develop the ability to break the bounds of Earth and fly, for air breathing creatures to extract Oxygen from sea water or for deep sea fishes to generate their own light? If these phenomena were not observable, a theory proposing such wonders would be hard to accept. But environmental variation and isolation are the primary factors which account for biological diversity. It is exactly because the Earth offers so many kinds of habitats that we have the extraordinary range of life that we do. We should not be surprised if stranger things still are to be found.

FLOATING CONTINENTS

The continental landmasses are not stationary structures that have been in existence since the beginning of the Earth as once was thought. They are not even firmly attached to the sphere of the Earth itself. They are in fact, broken parts of a relatively thin skin which covers only the outer portions of the globe and literally "float" on a worldwide sea of molten rock beneath. Not only are they relatively thin and not firmly affixed to the main mass of the globe, they are in constant motion as one would expect of anything which floats on a viscous liquid.

In the latter half of this century a revolution has been taking place within the sciences which investigate the origin and evolution of Earth. A revolution which, in many ways is similar to that which a hundred years earlier shook existing assumptions about the origins of man and the evolution of life. Although the notion of mankind as a changing, evolving species existed long before Charles Darwin, his *On the Origin of Species*, published in 1859 and *The Descent of Man*, published in 1871, which described the mechanisms by which evolution worked, created controversy that has still not entirely subsided even today. Although the principles outlined in these two works are now rejected by only a few of the most unenlightened, even well-educated and thoughtful people of the times were slow to fully accept, what now seem to be obvious conclusions.

The theory of continental drift when proposed was not entirely a new idea either, for others had observed the obvious; that the eastern coastlines of North and South America, seemed a perfect mirror image of the coastlines of western Europe and western Africa - as though they had once been joined. It was generally thought that this was mere coincidence, for there was no known geological process by which such a thing could be possible. As early as 1620 Frances Bacon had commented that these coastlines seemed to match perfectly, as though they were once joined. Benjamin Franklin, in 1770 speculated that planet Earth may be essentially liquid covered by a solid crust. Yet this idea remained only a fanciful notion until 1912 when a German meteorologist, Alfred Wegener, proposed that the continents were once part of a single landmass which he called Pangea (all lands), that broke apart and gradually moved to their present locations. He presented as evidence the fact that fossil remains unearthed on both sides of the Atlantic Ocean were of the same creatures, and that geologic evidence contained in rock strata on both continents also matched. Still the idea was not accepted. His proposal went even further, suggesting that the differences in flora and fauna indigenous to the continents resulted from isolated biological evolution that occurred after the landmasses separated.

Additional support for his theory came in the late 1950's when oceanographic data collected from around the world showed the existence of a continuous 46,000 mile undersea mountain ridge that ran through all the world's seas and wound between all the continents. How could such a feature, normally associated with landmasses, exist on the ocean floor?

Shortly after this, another clue to the changing nature of the Earth came from the discovery that magnetic poles reverse polarity about every half million years or so, and that lesser movements in the position of magnetic fields occur even more frequently. Why this happens is not known. It is assumed that currents in the molten outer core are in some way responsible. However, record of this change is recorded in molten rock flowing from undersea fissures which, when cooled by the icy water, forms mirror image pairs of lines of magnetism resulting from the separation which takes place as new ocean crust is pushed up from underneath. Molten rock permits molecular alignments which indicate the direction of magnetic fields at the time - as the lava cools, these alignments are permanently fixed. The fact that identical, but mirror-image bands exist, indicates that each band was split in the center as new crust pushed up from below and spread apart.

Here finally, was a model which might explain how continental displacement, of drift as it is now called, works: The pressure created by gravity pulling on the massive floating plates forces molten rock up from the magma (the liquid outer core) through cracks between the major plates on which the continents ride. Water pressure at these depths force the lava flow outward along the sea floor, cooling and hardening as it does. The spreading causes collisions at other places, forcing great sheaves of crust under the edges of adjoining plates. These so-called subduction zones are the deep ocean trenches - locations where ocean bottom crust disappears into the Earth forever. It is this constant creation and destruction of crust, taking place beneath the oceans, that creates forces which drive the continental plates to different positions.

Changes of this magnitude obviously take place over eons of time, far too long to be recorded by any historical documents. It has been learned only through deciphering clues left in nature that put together the pieces (literally) of the puzzle. It is now estimated that the time since the single continental landmass broke apart and the present is approximately 300 million years. This process is now firmly established as a major geological principle and has given rise to a new discipline: The study of the forces involved and the behavior of the continental plates as they drift around on the liquid outer core of the Earth is called Plate Tectonics.

The final acceptance of Wegener's theory has taken more than 50 years and has been a revolution in science on par with the acceptance of the Copernican model of the solar system and Einstein's relativistic model of the universe. Revolutions in thinking which cause old theories to be discarded or drastically revised are understandably slow to be accepted. It's easy to understand how movements of the magnitude of continental drift are difficult to comprehend. It is only when new clues defy explanation by old theories that more accurate and inclusive models have to be acknowledged.

The significance of continental drift is not only geologic - it is biologic as well, for whatever shapes the environment shapes the course of evolution. The division and drift of the continents which caused the climate and terrain of each to differ so markedly, profoundly affected plant growth and animal life, creating new habitats and cradles of life. Biological isolation also meant that similar species could no longer interbreed and would eventually diverge. It was, as we shall see in later chapters, changes in environment, and biological isolation that precipitated the divergence of humans from the apes.

It should be obvious that forces which can move not only mountains but whole continents, are responsible for other monumental upheavals and changes upon the land. Earthquakes and volcanoes both are products of the friction created as plates move against one another or when one is forced under another: In the case of earthquakes, the edgeways sliding of plates, or fractures in plates, cause the tremors which take place all the time and generally go unnoticed except by sensitive seismic measuring devices, but do occasionally cause such destruction. Volcanoes are generated along and back from the edges of subduction zones, where heat is generated from the friction caused by one plate being forced under another, melting the rock which is then forced out elsewhere as lava. Mountains are also caused by the collision of continental plates, but instead of subduction, the mass simply crumples, rising higher into the atmosphere creating the majestic features that decorate every continent. For this reason, it is not simply chance that most great ranges occur nearer the outer edges (or what was once the outer edges) of continental boundaries.

The separation of the single original continent, Pangea, into separate continents, the formations of the world's great mountain ranges and the establishment of present-day seas all occurred long before human beings existed. When these events took place, our ancestors were still in trees - indistinguishable from apes. However the changes did establish irregular features which allowed the full potential for biological diversity to create extraordinarily varied and distinctive adaptations to every habitat.

Although human beings came on the scene only in the last two million years or less, long after the continents were no longer one, still have managed to migrate all over the face of the Earth facilitated through geological changes brought about by another set of forces. And as we shall see, even more extraordinary manifestations of evolution may have taken place as a result.

ICE AGES

Continental Drift is not the only powerful force at work which shapes our planet and forces major life adaptations. Seven times in the last 1.6 million years, approximately every 100,000 - 200,000 years or so, glaciers have covered most of the continents in the Northern Hemisphere with thick seas of solid ice, sometimes as thick as two miles. This Pleistocene epoch was marked by wide fluctuations in temperatures which caused four major and three lesser ice ages. The reasons for the glacial advances and subsequent slow melting are not completely understood. Certainly the global temperature had to vary sufficiently for this to occur, but why this happened is not clear. Various theories have taken into account geologic, atmospheric and astronomical activity, such as volcanic eruptions, increases in Carbon Dioxide in the atmosphere, variations in the Earth's position relative to the sun, interstellar dust clouds or changes in the sun's energy output.

The lasting effects on the surface of the Earth remain however, despite the considerable erosion and weathering that has taken place since. The movement of these massive oceans of ice caused enormous crushing, gouging and buckling in the Earth's surface that is equaled only by the effects of earthquakes and volcanoes - and has affected most of the continental surface of the Northern Hemisphere. During the most recent period of glaciation, about 20,000 years ago, ice flows on the North American Continent reached as far south as Kansas and Missouri and in Europe as far as London.

During the thousands of years required for the slow advance of these glaciers, animals and humans (for Homo sapiens, not very different from ourselves, and their direct predecessors Homo erectus, were around during these times) sought shelter from the unrelenting bitter cold. Caves, then as now, provided the best possible natural shelter. Throughout the world these cave-dwellers have left the evidence of their occupation: tools, weapons, campfires, artwork, personal items and most importantly, skeletal remains. Many of the most significant archaeological finds, yielding invaluable information about early human development, have been located in caves.

It is easy to imagine these humans gathering within the relative shelter and safety of whatever openings in the Earth's surface that were available. They built fires near the openings for warmth and to ward off dangerous animals. They established small settlements and may have ventured out only to forage and hunt. As the climate became increasingly harsh and food supplies dwindled, many of them perished, either from starvation, exposure or both. Others, possibly most, slowly migrated southward in pursuit of game animals and vegetation. Of course people of this period would have had no way of knowing that warmer weather was to be found farther south. They simple moved in the direction that food seemed most plentiful.

To be sure, this was a very slow process, requiring thousands of years, and without written histories and a means of recording and communicating over time beyond one generation to the next, to these people life seemed normal at any given period and they would have had no awareness of a long-term climatological threat.

In time, the caverns and crevices in which early man huddled from the cold were closed over; sealed with millions of tons of solid ice, perhaps most still completely unknown to us, for the areas affected by the most intense glaciation are still savagely cold, unsettled or only sparsely so, and certainly less investigated by explorers, studied by scientists, and scrutinized by developers than more temperate regions. The most startling discoveries about the evolution of mankind probably remain to be found. Perhaps deep under the cover of present-day glaciers, ice fields or permafrost, lie more homes of our Pleistocene ancestors.

Most anthropologists hold that man simply migrated to evade the ice and that those that did not died out. After all, evidence suggests that significant migrations took place from continent to continent when land bridges such as that which connected North America and Asia (in the region of the Aleutian Islands near Alaska) made this possible. However, changes in weather and climate can be an intense motivator and would result in a more desperate effort to gain immediate relief than mere slow migration would afford.

Current views of the Pleistocene period are that the climatic changes were more sudden that previously thought and, by contemporary standards, would be thought of as a climatic catastrophe. And even if the actual advance of the glaciers was slow, arctic cold fronts, which must surely have been more severe then than now, and could have descended suddenly and with increasing rapidity, would have made finding shelter a much more urgent need than migration.

These early humans would have had no cognitive understanding or cause even to believe that movement to another place would result in relief. For all they knew, it was cold everywhere - the entire world was the same - flight would have been pointless. There are many cases even today, of intelligent, educated people remaining in regions which hold impending danger despite the knowledge that fleeing would be advantageous. Reason suggests that under the conditions of severe cold, whatever shelter within the comparative warmth of the Earth that could be found would be used, and for as long as possible.

Those northernmost populations most threatened by severe cold were not the only ones to use caves as shelters. In even the most temperate zones: Africa, South America and Australia, humans also occupied natural enclosures. In these regions, not directly affected byglaciers, the ice ages still had a profound effect: As the planet grew colder and more of the world's water froze, the level of the seas dropped, exposing vast areas of the continental slopes as new terrain for human discovery. It was this process which created the land bridge between Asia and North America and several others between major continental masses which permitted the distribution of humans to all continents. Any openings found in these areas surely would also have been used by migrating peoples. The melting glaciers released the water, raising sea level and cutting off the areas, submerging, once again, the cave sites. This cycle was repeated many times in Earth's history and affected every landmass, either through the direct impact of glaciers or the indirect impact of fluctuating sea levels.

The significance of the Pleistocene epoch was that it occurred at so critical a time; exactly during the development and spread of early man - the time during which the symbiotic bond between humans and the Earth was never more intimate. The climatic upheavals which motivated the migration of people over land may have also caused their migration inward. Perhaps the seeds of life were spread not only over all the landmasses, but also planted beneath the surface.

Extreme climatic changes are known to coincide with periods of intense evolutionary change. New species appear and old ones die out with greater frequency than in long periods of relative climatic stability. One possible causal connection between these occurrences is an obvious one; that changes in environments which create extreme conditions enforce adaptive natural selection more rigorously than more moderate conditions. Another factor may be at work also. Some theories propose that environmental changes cause separation of groups from one another, and that it is this separation that allows greater diversity to develop. As once homogeneous populations are fragmented, isolated in different habitats, new species arise, some of which survive while others become extinct.

It may have been just such a process that was responsible for the earliest beginnings of mankind. One theory holds that it was climatic changes which stimulated the divergence of man from the apes and began the long process of development into present-day humans: Around eight to nine million years ago, in the area of East Africa around the East African Rift Valley System, an area which had been a lush tropical region, began to dry out as crustal plate movements changed the landscape from a relatively flat jungle to a varied terrain with smaller islands of jungles surrounded by large grasslands with little rain and limited vegetation. It was at this time, apparently because of the juxtaposition of such varied terrain, that some of the tree-climbing apes were forced out of the jungles and assumed an upright stance in order to survive by scavenging and hunting on the open plains. This was the origin of man. Had it not been for the climatological changes which forced these adaptations, the evolution of humankind would likely have been strikingly different.

The more recent evolution of humans is by no means a simple lineage defined by a clear trail of direct ancestors. The reverse is true. As more fossils are unearthed and studied, the picture grows more complicated. About 2.5 million years ago precursors of modern humans existed in considerable variety and some authorities now believe that as many as 16 different Homo species may have existed at various times during the last five million years. Where we fit into all of this is more murky now than ever.

One thing is common to virtually all early humans though; they used caves, not only as shelters but as burial sites and locations for rituals and ceremonies. Judging from the artifacts discovered in caves throughout the world, even the earliest peoples had a reverential feeling for the underground and ornamented the walls and roofs with murals depicting all kinds of creatures in various actions, from mating to life and death struggles with other animals and even human hunters. The purpose of these rather elaborate drawings is a matter of considerable speculation among anthropologists. Many feel that they represent a semi-religious or superstitious practice intended to insure successful outcomes to imminent hunting expeditions. Others believe that the artwork is just that; personal expressions of the artist's view of the world.

Whatever the reason, caves were the chosen spots for their activities, despite the fact that these were the most difficult locations imaginable for such work. The Lascaux cave in southwest France, discovered in 1940, contains carefully drawn color murals, not just near the entrance but throughout, as far as it is possible to penetrate. This is not atypical: Artwork found in the Niaux cave and the Trois Freres cave in the French Pyrenees also contain artwork in the deepest recesses. The extent to which caves have been utilized by early humans, the various artifacts discovered and their significance in the progress of humankind is a subject unto itself and would make for several books. The point is that caves were an integral part of life for the earliest humans and even their precursors. Even today, many religious groups hold caves in reverential esteem, particularly in the East, where many caves are still considered shrines by Hindus.

It is easy to imagine the appeal caves had for ice age people. Not only did they provide shelter from the wind, rain and snow, they were also places of relative safety which could be made secure simply by building a fire in the entrance. Some heat would be retained in the enclosure, making it far more comfortable than any other shelter that could be constructed and burning shafts of wood from the fire would made convenient handheld torches for explorations deeper into the interior. Most known caves show evidence of incursions into the remotest parts, limited only by the dimensions of the interiors themselves.

THE THAW

We have no way of knowing how many or how extensive such earth shelters might have been. Hundreds of enormous caverns which we now can access, most discovered comparatively recently in our history (by modern humans at least), all lying relatively shallow beneath the Earth's surface, give ample evidence to the potential massiveness of underground bulwarks. How much subterranean space is yet to be found deeper within the Earth or under areas still covered with ice or water?

The most common type of cave found throughout the world is formed by a process of limestone erosion. Water, flowing downward from the surface and through lateral underground fissures, dissolves hard limestone and slowly expands these cracks into larger openings. In some cases, underground streams or rivers carry sand and other grit at greater velocities which wears as well as dissolves the rock, accelerating the formation of openings. This is not the only process of course, there are others, including volcanic activity, but limestone dissolution is the most common.

On the surface, the entrances to unknown caves are sometimes obscured by foliage and discovered by only by accident. But entrances are also often revealed by weathering processes, making their discovery more likely. Caves with openings underwater however, may be more difficult to find; the rapid accumulation of sediment would quickly bury all but the largest, concealing many unknown entrances before the quite recent techniques of underwater exploration would have allowed their discovery. Doubtless, there are many caves, especially underwater, about which we have no hint.

Caves about which we know are largely an upper crust phenomenon. Since they are formed by water flowing from the surface, their superficial openings usually lead to chambers located only a few hundred or a few thousand feet down. The greatest numbers of caves tend to be in highly-populated areas; probably due to the fact that their discovery is often the accidental by-product of other activities. Uninhabited, sparsely populated or undeveloped areas surely have many yet-to-be-found openings. Deeper caves or caves which may have formed offshore, in areas that were once exposed but that are now covered by oceans and silt, will be harder to locate.

Earth scientists have a basic model of the various layers beneath the Earth's crust that is reasonable reliable. Seismic analysis, among other sources of data, yield good information about the density and approximate temperature of the Earth all the way down to the core. However this is only in crude generalities, and the greater the depth, the less clear the picture. The demarcations between layers are defined in hundreds of feet and tens or even hundreds of miles. Of course much of our information about the immediate depths comes from direct exploration and drilling. However, relative little of the actual surface land area and far less of the ocean floor has been penetrated by deep drilling. It is quite possible that vast crevices exist in areas yet explored - areas too shallow to be significant in the measurement techniques now at our disposal. It has only been in the last few decades, through the use of the most modern and sophisticated methods, aided by satellites, that measurements exact within a few hundred feet have been made of the surface. Far, far less is known for certain about the underworld.

For many years it was the general assumption among Earth scientists that despite the variations found in the upper crust, the Earth's lower crust was relatively uniform. That assumption is now doubtful: Seismic reflection profiling - analysis of echoes from intentionally generated waves created by surface explosions - has revealed an unimagined degree of complexity. It now appears that even in areas with few distinctive surface features, the entire underground landscape contains an extraordinary degree of irregularity. Seismic reflection has been used for many decades by the petroleum industry to locate oil deposits. Since the 1970's more purely academic studies have been made and show that the crust, all the way down to the mantle, contains far more folds, faults and variations in density that imagined. Maps have been created which show these irregularities, but the actual significance of the profiles is unclear.

Another assumption that has proven to be false is that the crust beneath mountains is as thin as in non-mountainous terrain. In actuality, mountains have roots that lie as far or farther beneath ground level as the mountain rises above. Beneath large mountain ranges such as the Himalayas, the Earth's mantle is deflected downward beneath the mass; the mountains actually float on the mantle just as an iceberg floats on water, with most of its mass unseen. The crust and its underlying mantle function (on a massive scale) as an extremely dense liquid. This principle of isostasy accounts for another phenomenon related to the melting of glaciers. In the region known as Fennoscandia - Norway, Denmark, Sweden and Finland, the massive weight of ice during the peak glacial buildup displaced the ground underneath by several hundred feet. After receding, the Earth slowly flowed back to its original level, or nearly so, at a rate of about 2 centimeters per year.

This is still going on and has for the last 5,000 years.

That primitive man used caves as shelters is not a matter of speculation. Neanderthal Man is so named for the caves in the Neander Valley near Dusseldorf, Germany where the remains were first located. Peking Man, another landmark find, was also discovered in caves accidentally unearthed during quarrying operations near Peking (now Beijing), China. It has been generally considered that early man dwelt near the openings of caves and used them only as temporary refuge from the elements. However, the increasingly bitter cold of the climate likely would have driven them deeper into the comparative warmth of the caverns. Many caves in France and Spain show evidence of such deeper explorations. And although caves are generally very sparsely populated by any kind of life, making them unsuitable for long-term continuous habitation, in caves that may have been opened by receding seas, there may have remained pools of water which contained trapped marine life, as is often the case when floodwaters recede.

This would have provided a convenient source of food for scavenging and made cave occupation much more attractive.

Certainly with more and more of the area covered with ice, less food from both plants and animals would have been available. It is possible that our ice-age ancestors, migrating deep into surface openings, stumbled onto connections to larger spaces and were unable to trace their way back, or if they eventually did, found the original point of entry forever sealed by snow and ice. Many other openings in the Earth would have been sealed not only by ice, but by the rising seas; land areas which would have become occupied by humans when worldwide temperatures dropped would have slowly flooded as the reverse occurred.

The best evidence now indicates that humans spread from their origins in Africa to the rest of the world over a period of two million years. During that migration the level of the seas rose and fell many times with each ice age. Since we know that early humans inhabited some areas now under water but which were dry land at times of peak glaciation, and since it is well documented that these early people also used caves as shelters, we can reasonable assume that any openings they found in these periodically exposed areas would also have been used.

The Earth's crust is much thicker under the continental masses than under the oceans - quite a bit thicker. On average the continental crusts, which range from about 12 to 45 miles thick, are about four times the thickness as the ocean floor crust, which averages only about 6 miles. The transition between the continents proper and the sea floor is comprised of the continental shelf, the submerged area which forms a sloping platform outward under the sea, and the continental rise, the gentler final slope which extends all the way to the ocean basin. Together these zones expand the actual total area of the continents by about 15 percent and may extend offshore from hundreds to thousands of miles. They are more expansive in the Northern Hemisphere and even more so on the northernmost sides of Europe, Asia and North America - the areas most directly affected by glaciers.

Any caves or caverns which exist in these transition zones or extend from them deeper into the continental plates are likely be nearer to the mantle than those which exist on land. For this reason alone, the probability is that some feature geothermal springs or exposed magma. If early humans were to stumble upon caves with such sources of heat, it would have been a discovery equal or greater in magnitude than the discovery of fire. In a severely cold climate (and the times when these caves would be accessible would be during the coldest periods of the glacial cycles) the discovery of constant sources of heat would have been welcomed indeed. The discoverers would surely have returned again and again to warm themselves, perhaps cook food, make weapons or even bathe, if heated water at some locations was not too hot.

Any substantial amount of exposed molten rock would not only provide heat but light as well, at least as much light as a small fire, and once eyes became used to the low level of illumination, would be adequate for activity. Any such warm oasis would have been far more comfortable then the barren, bitter cold of the outer world.

Sources of food from within may have been located and trips outward to hunt would not always be needed. The food may have been fungal plants or sea life of some kind, if the caves opened to the ocean. Harvesting food from pools of water or scavenging for fungal plants would be at least as attractive as the dangerous business of stalking large carnivorous land animals. It's possible that, just as agriculture finally supplanted hunting as a means of providing food on the surface, these cave dwellers may have learned primitive aquaculture or subterranean agriculture.

Geothermal sites would have been warm, nurturing places that might have caused early humans to revere the inner world, to think of it as more attractive than the bitter cold, and thus minimize their outside activity. Some may have existed for thousands of years in a dual existence; living partly within and partly without their new homes.

Since we know that humans sought shelter in caves, and if it is reasonable to assume that under increasingly harsh conditions they retreated deeper and deeper, using these shelters to the greatest extent, it is also reasonable to assume that many became hopelessly lost, having ventured so far beyond their original point of entry that they were unable to return to the outer world, or if they did, found their passage blocked by avalanche or accumulations of snow and ice, to be forever trapped. Or did some find the environment so preferable to the outside, that they simple chose not to return? Did some groups split off from others to follow a life completely underground? The question arises; if this is the case, did any survive, and if so, for how long?

Is it possible that, just as long-thought extinct specimens of life from hundreds of millions of years ago still live in the oceans, some of our ancestors have managed to survive somehow, underneath the Earth? Is it even possible that, just as the exceptionally adverse conditions in the deepest ocean trenches have enforced adaptive evolution into incredibly unusual forms of life, underground conditions have created a most unusual race of humanoids? Trapped and isolated from the environmental and biological influences of the surface world, could survivors have undergone an evolutionary process that diverged entirely from ours? Do they still exist?

Could It Be

Of the requirements for sustaining life: air, food, water and shelter, the last is a given in a cave environment. Water also may not necessarily have been a problem. Underground streams or seas, possibly even water seeping from above could have provided an adequate supply in some areas. Water exists in abundance in most known caves and there is no reason not to believe that caves even deeper than any yet discovered will also have water. Light is not essential for life, except as a source of energy for plant growth, and ultimately, animal protein. However abysmal those caves may have been, the inhabitants would not have perished from the dark. And we have seen from the example of life forms now existing at the deepest bottoms of the oceans, where no light penetrates, that absence of sunlight is not a barrier to survival.

Many of the life forms discovered at the depths of the oceans survive on a food chain that has links to the surface, and therefore indirectly to the sun. In these cases, small creatures feed on the decaying organic matter that slowly drifts down from nearer the surface, and in turn they provide food for larger creatures in the more or less conventional food chain with which we are familiar. This dependence on the slow accumulation of organic materials from the surface is the limiting factor on the population at these depths, which is much sparser than nearer the surface. However, a most startling method of biological energy conversion has been recently discovered on the Galapagos Rift in the area of the East Pacific Rise.

Strange tube worms as long as twelve feet can be found in large clusters, as well as giant clams five feet long and giant crabs living on the ocean floor over two miles down. The clams reproduce 500 times faster than related species of shallower waters and have red flesh which is rich in Hemoglobin that allows them to thrive in the low-Oxygen environment. These creatures survive not on organic material from the surface, but are adapted to a food chain that is completely independent of energy from the surface. In these areas, lava from beneath the Earth's crust flows up through fissures in the ocean floor and reacts with the cold water to acquires sulfate ions and cause the water to become acidic. Bacteria metabolize the sulfur compounds in the hydrothermal water and provide a source of food for the larger animals in a food chain that is completely separate from the surface world and the energy provided by the sun.

In this environment, the energy comes from the heat of nuclear decay taking place within the interior of the Earth and not the nuclear reactions within the sun. These undersea oases feature chemosynthesis as opposed to photosynthesis as the primary source of energy for flora and fauna. So efficient is this conversion, that some scientists have speculated that sites such as these may have been the original sources of life on Earth, although this is now not generally thought to be the case. Many of the creatures discovered in these locations are so unlike any others on Earth that new species, genera and families had to be created for their classification. In fact, the tube worms are considered so unusual that they have been classified as an entirely new phylum.

So only air remains as a barrier to survival in our speculative thinking. Sea water contains considerable amounts of Oxygen - enough Oxygen to support enormous forms of life at very high levels of aerobic activity. Sharks, for example, have gills different from other fish, and cannot remain still for long. Instead of forcing water through by muscular motion, sharks rely on the flow of water forced through the gills by the swimming motion itself. This has the advantage of increasing Oxygen extraction as energy needs increase. Extraction of Oxygen from seawater is by no means too inefficient to support large, intensely active creatures, and in large numbers too, if we look at the abundance of fish in some areas.

This is not to suggest that underground humanoids have developed gills, but that the Oxygen is there and that a natural process may provide an Oxygen atmosphere indirectly from sea water: Boiling water releases its captive Oxygen. We are all familiar with natural geysers such as Old Faithful at Yellowstone National Park in Wyoming. These erupt regularly when underground tidal motions cause water to come in contact with hot portions of the mantle. If something similar occurs but is not vented to the surface as steam, but rather boils steadily or at recurring intervals, the liberated Oxygen would be contained underground and create a breathable atmosphere in any open chambers and connecting spaces. This might occur as tidal influences change the distribution of underground water, bringing it into contact with adequate sources of heat, or vice-versa; gravitational forces might cause slight changes in the position of magma and redistribute heat to intermittently contact underground water. In this respect, the underground atmosphere may depend on the cyclical changes in gravitational forces just as our food supply depends on the vagaries of season, sunlight, warmth and weather.

There may be only a few actual locations where these unique circumstances would simultaneously occur to support life - but as we shall see in the next chapter, there may be a lot of room underground.

It is established that life deep underground, cut off from the surface world is entirely possible. In fact, such life forms have long been known to exist within the caves that have so far been studied. Troglobites are species of creatures that have become so completely adapted to life underground that they have evolved entirely separate from surface species. They are the common dwellers of caves known to us: bats, flatworms, various fishes, salamanders, crayfish, cockroaches, insects, rodents and bacteria. But they are forms of these creatures that are significantly different form the ones found on the surface.

These known, exclusively underground creatures are dependent on surface life by virtue of the fact that organic debris drain into the caves and provide a source of food for simpler forms of life which in turn support higher species, just as has been observed in the case of most deep ocean life. But as we have seen, this need not be so - and the ocean depths may well be considered far more inhospitable to life than caves, in that the tremendous pressures that prevail in the oceans are not present to create additional hardships to life underground. It is possible that mechanisms within deeper caves, those that are closer to the mantle, are sources of energy and chemosynthesized food supplies that are not dependent in any way upon the sun.

The Earth itself is a tremendous store of energy, both from the radioactive decay of unstable elements deep underground and from indirectly stored energy from the sun in the form of oil, coal and gas deposits which are nothing more than decomposed organic matter that was once of surface origin. Just as there are now known to be metabolic processes in some bacteria which can sustain life from the sulfate compounds in seawater which has reacted to molten rock, so may there also be similar interactions which occur underground that feed lower organisms and create a food chain which may support more developed underground species.

Bacteria which thrive on petroleum have been developed to degrade oil spills in order to minimize ecological disasters. It may be that similar microorganisms form a vital link in an underground food chain that converts oil or coal fossil fuels to animal protein - completing circular process that began when prehistoric flora and fauna decayed to create these deposits in the first place. This would not be unlike a process that takes place in the ocean depths, in which bacteria live in the decaying sediment on the sea floor and indirectly support higher forms of life. In fact, if bacteria can flourish in the geothermal springs on the ocean ridges, then any deep caves that hold bodies of water and may be close enough to the mantle for magma to flow, may also have identical chemosynthetic "fountains of life" at depths shallow enough to permit harvesting.

Is it possible that deep underground fissures have unique environments which can harbor and nourish life, making them terrestrial analogues to the mid-oceanic ridges or the deep ocean trenches? Given the development of such diverse and incredible forms of life as those found in such locations and their discovery in recent years, it should not be surprising if, in the future, the Earth reveals even more remarkable creatures. In the most literal sense, we have only scratched the surface in terms of exploring our world.

The prospect of finding intelligent life on other planets or even other galaxies has now been given serious consideration and is no longer treated as pure fantasy. While there has been increased interest in the prospect of discovering intelligent life on extraterrestrial worlds, serious investigation of the possibility of intelligent life from deeper within our own planet has been largely ignored.

Room Without a View

If it is theoretically possible for higher animals, even humanoids, to remain alive entirely within caves, we have to consider if there is sufficient potential space within the crust of the Earth for substantial populations to have developed and evolved, for it is within the upper levels of the crust that life would be limited. The idea, often illustrated in science fiction, that there may be extensive caverns or tunnels which extend deep into the Earth's interior, plainly cannot be so. The crust ranges in thickness from a few miles to 25 miles or so, and the temperature rises sharply with depth, reaching 700 degrees Fahrenheit within several miles of the surface. By the time the upper mantle is reached, temperatures are 1,400 degrees F. The existence of life more than a few miles deep would not be possible. Pressure as well as temperature rises and within a few miles of the surface, would be so great as to compact any cavities.

But let's consider how much space that is. Even if we restrict the possibility of habitats for potential life to within one mile of the surface, considerable space remains available for extensive caverns that are completely unknown to us. In fact, the greatest areas of the Earth's surface under which hollow spaces may occur are not easily explored, and account for the major portion of the world's exterior. The oceans, seas and various bodies of water cover better than 70% of this area and ice caps which still remain, cover another 10%, rendering even more of the remaining land area virtually uninhabitable. So far as we know, oceans of the magnitude that cover the surface do not exist underground, nor do ice fields, and so, theoretically, a relatively larger portion of the underground is available for potential occupancy than the surface.

There may be even more room than these conservative estimates allow. The crust thickness varies considerably and is as much as 40 miles thick in some places underneath continental plates. The sharp rise in temperature with increasing depth is also quite variable. Under central continental regions where the surface temperatures are and have been cold for a long time, the rate of increase may be as little as 1 degree Celsius or less per hundred yards. This would mean that, in terms of temperature, sites as far down as three to five miles might be found, if structural formations supported cavities.

That the Earth is uneven in density and may have many unknown spaces is evidenced by variations in gravity from place to place on the surface. The Bouguer anomaly is the term for variations in the effect of gravity on theoretical sea level around the world. Satellite measurements taken by a European Space Agency satellite indicate that sea level (and this includes land areas which are defined as being at sea level) vary as much as 350 feet. This variation is not simply the tidal effect, but represent actual variations in the force of gravity as measured in space. By definition, variations in gravity are the result of underground variations in density, which points to an extremely complex underground geography, the likes of which had not been imagined only a few years ago.

We humans occupy quite a small percentage of the available space on the Earth's surface. If we further consider those vast areas which still remain very sparsely populated, that percentage is reduced even more. We have expanded to use the space above us, but only extremely recently in our history. We now routinely use the space that is far above us, but that much "headroom" is certainly not essential for life. Until only a hundred years or so ago, prior to air travel, we occupied only the very surface of the globe and, excluding tall buildings, required space not very much above our actual height. Humans could live quite successfully in the shallow spaces that might be within a few miles of the surface. In fact, throughout most of our history we have developed and progressed utilizing, in absolute global terms, a small fraction the space on the surface.

We cannot assume that for life to continue and even for a civilization to develop, large open areas would be necessary. The will to survive is strong and people and animals function with what is given them. If only a maze of catacomb-like spaces, hardly larger that the occupants themselves were what was available, they would find ways of adapting. Even if only one unknown cave exists that is the magnitude of Mammoth cave in Kentucky, that alone would be sufficient to house a substantial population, a population large enough for civilization to develop. Mammoth cave consists of 510 miles of connecting corridors spread over many layers with several large chambers. A cave with comparable space could easily house several hundred thousand inhabitants. This seems impossible for us to imagine living under such conditions, but if born to it, and no other environment were ever experienced, it would seem the norm.

Furthermore, an underground people would not necessarily be limited to the naturally occurring space in which they evolved. We have expanded the area that we can inhabit as our population has increased and so might they. Our methods of reclaiming additional land have traditionally been deforestation, landfills and drainage. As the need for ever more space grows with our population, more aggressive and technically advanced methods will have to be undertaken to continue this trend. This and finding ways of safely disposing of the waste we generate are some of our greatest contemporary challenges.

Underground, methods of expanding the available space would require draining water or oil-filled areas and excavating solid materials. The difficulty with removal of material, either liquid or solid, lies in where to put the unwanted waste. Drainage would require that other locations which are lower in elevation, and are too irregular or otherwise unsuitable for habitation, be used to contain the transferred liquids. Solid removal may be incorporated in reshaping the environment and may even play a part in their technology, which will be discussed in following chapters.

Digging is one of the most basic kinds of work and for an underground species would be almost as natural as walking. The earliest use of tools in such an environment might not be axes to chop and whittle wood, but stone implements with which to dig and carve rock. Long before any higher activities developed, perhaps even as long ago as a million or a million and a half years ago, a burgeoning underground species may have began to explore and expand their environment by tunneling. Eventually, some success may have led to the discovery of connecting spaces, and continual exploration became as much a way of life for them as it has historically been for us. Their explorations also may have resulted in more significant and startling revelations as will be discussed.

THE LOOK OF LIFE

The strength of evolution lies in biological diversity and the perpetuation of adaptive traits. Species which cannot successfully adapt die out and those which can, become increasingly well-suited to their environments. This process of natural selection occurs in humans as well as in other living things. Populations which have established themselves in certain geographical regions bear distinctive biological traits we can recognize as adaptive assets for their native climates. Although these are mostly subtle, skin pigmentation as a mechanism for protection from the sun is an obvious one, for the sun is one of the elements most hostile to life on the Earth's surface. Human beings forced to adapt to a drastically different environment, one entirely without sunlight, in time would exhibit features quite unlike our current appearance.

Troglobites, cave lifeforms which we know to exist, differ markedly from their surface relatives. All are blind, their optic faculties having atrophied to such a degree as to be genetically established.

In most species, the actual eye is absent or is vestigial. Also, vision centers in the brain are much smaller than in surface relatives. Ears are more prominent and feature refined balance mechanisms which do not rely to any degree on vision for orientation. Posture and balance centers in the brain of underground species are also much larger. In addition, odor receptors and vibration reception centers are much more developed in subterranean creatures than in surface dwellers. These are all features one might well expect if natural selection through genetic mutation altered physiological traits over time to adapt to life in complete darkness. Pigmentation, a trait essential for camouflage, is also useless in total darkness, and so as we might expect, is subdued of absent in most cave species. Physical size is also somewhat smaller in most cave species and most are metabolically slower than their surface relatives. Cave creatures must get along with reduced food supplies and need to be more fuel efficient.

The traits which so enable cave dwellers to exist in complete darkness would prove to be serious or fatal liabilities in the active, competitive world of the surface. Most would be easy prey for larger carnivores. Their lack of vision, slower movement, lack of camouflage and competitive hunting skills would doom them to early extinction outside their normal environments.

In considering the possibility of troglobitic humanoid evolution, the most serious obstacle is the food supply. Most known cave environments do not support larger carnivores or any kind of larger warm-blooded animals. The metabolic requirements of such creatures demand an amount of nourishment that is unavailable in such a closed food web. However, most of our current knowledge of cave life is based on sites that are accessible from land areas. A great many contain water and fish, though not in the abundance of surface bodies of water. Of those that contain lakes or streams, most are freshwater, their source being of continental surface origins.

It may well be that the most extensive cave formations on the planet are accessible only from under the seas and that their sources of water may be the saltwater of the oceans themselves. If this is the case, there may be an adequate supply of marine life that circulates freely to and from the caves, and that larger mammals have had ample food for evolution. Such interior inhabitants would be trapped, not beneath the Earth proper, but beneath the depths of the sea.

What are the implications of species adaptation to life underground for possible humanoid deviations from our current form? From the differences between underground and above-ground specimens studied, and from the changes we can imagine that might be required, we can make some educated projections: Physical size would probably be less, both in weight and height, probably from the reduced calorie intake (if we assume that food in the abundance and variety of surface supplies is not available). Musculature would also be less pronounced (as a consequence of lowered metabolic activity and possibly a difference in growth hormones from the lack of sunlight - also from reduced physical activity necessary for survival in an environment without changes in weather). Of course pigmentation would be lacking and skin would seem translucent. With body size reduced and brain size presumably not, the head would seem much larger in relation to the body. As a consequence of uniformly mild or warm temperatures and no weather as we know it, hair would probably be very sparse or absent, both on the body and head, and with no eyebrows (there being no need to shield the eyes from the sun's harsh glare). The general appearance would be one of rather slight, pale and fragile creatures.

Ears would be extremely large and flared or might be entirely absent, having mutated into vibration sensors on the sides of the head which do not protrude at all. Localization of sounds depends on phase differences between two ears, and sounds must be of the shorter wavelengths for best directional distinction. Lower frequency sounds radiate in broader patterns and are more difficult to localize. In an underground environment, a multiplex of reverberations would make detecting the exact location of sounds difficult or impossible. Therefore lower frequency vibrations, which are well transmitted through the Earth might be more useful in indicating movement in the environment, therefore the advantage of directional, amplifying flanges around the ears would be lost. Additionally, without sight or with limited vision, distance recognition and perception of sound from all directions, not just from the direction of the field of vision would be especially important.

Eyes may be either nonexistent or nonfunctional vestiges. However, if adaptation is not complete, if some source of dim light such as chemical light exists, or if they have mutated to become sensitive to radiations of other spectra, the eyes might be extremely large. For example, one plausible possibility is that is that visual sensory organs could have evolved into receptors for longer wavelengths such as infrared. Recent technological development of night-vision glasses and binoculars has demonstrated how effective such different modes of detection can be in subdued and total darkness.

Infrared wavelengths are emitted by sources of heat and visual images are formed by differences in the amount of heat contained in the environment. Certainly such a manner of seeing would be effective for finding warm-blooded prey. In fact, such sensory capability would be an even greater asset since it not only allows the detection of warm-blooded creatures in the dark, but also can show residual heat images from locations that they have recently occupied, making infrared vision a distinct advantage in stalking and tracking. Infrared waves can also be useful in distance perception, as anyone who has used an autofocus camera can attest. It can even be used in communication (as with audio and video remote-control devices). Creatures with such vision would be as unable to see in our world as we are in the dark, and would require some sort of filters or goggles to modulate visible light just as we must have infrared goggles to see in the dark.

Speech or vocalizations could have become a necessary part of navigation and location sensing as well as communication. Therefore the pitch would be significantly higher than ours, perhaps as much as several octaves, and would likely be of short duration, since audio ranging systems, both man-made and natural, rely on short bursts of high-frequency sound. So refined is the echolocation system of bats that they are able to navigate through caves relying entirely on it, not only to avoid flying headlong into walls or stalactites, but also to hunt small flying insects, which they must do at a prodigious rate to satisfy their warm-blooded metabolisms. Bats navigate so precisely that they are able to avoid an obstacle as small as a fine wire or can fly through a small hole in a net stretched out in their path. Bats are also extremely efficient hunters. Studies of bats have shown that, on average, they pursue, eight insects per minute and catch six of them. By audio ranging alone, they are able to catch enough flying insects to completely fill their stomachs within one to two hours - better than many predatory animals with vision can do.

One can imagine that creatures with some form of infrared vision, coupled with a keen sense of smell and some form of high frequency sensory perception would be even more effective hunters than most surface animals, as they would have to be in an environment with less abundant food supplies.

The noses of our human troglobite may be more prominent in some way. They may not necessarily be longer, but may have larger, flared nostrils for enhanced smell, giving the appearance of a pug, turned-up nose.

Hands and feet may be slightly larger, but with somewhat longer fingers and prehensile toes to aid in climbing walls and ambulating over uneven terrain. Arms may be longer than ours relative to their bodies, since this would also be a decided advantage.

Gender differences in subterranean humanoids would remain, but distinctions based on physical size would not be evident. Without the physical requirements needed to compete and hunt large land animals over vast distances in inclement weather, male physical dominance would have disappeared. Copulation and reproduction would likely be little different but gestation may take far longer, due to fact that nutritional requirements of a developing fetus would place a strain on already limited supplies.

Patterns of food availability may necessitate very different eating habits from our own. Cave environments which are now known feature extremely efficient food webs. Almost nothing is wasted. As already mentioned, most caves get some organic material from the surface introduced into the system, washed there by surface water seeping into ground fissures, or from streams which actually meander along and below the ground . It may be that if some caves are open to seawater, organic material is deposited by seasonal or tidal variations and either provide sources of food directly or nourish the growth of flora that support higher organisms. Inhabitants of the Nile Valley, until very recently, were entirely at the mercy of the seasonal flooding of the Nile River to provide growing conditions for grain. Despite this limitation, they developed a flourishing civilization under what would otherwise have been impossibly harsh desert conditions.

Whether biological material is obtained from geothermal sources, oil deposits or from sea organisms, the cycle of distribution would certainly differ greatly from our seasonal, agricultural growth which depends on weather and solar proximity. It may be that extensive hibernation has become a part of the underground life cycle and may even feature ritual cannibalistic behavior. Certainly the early period of survival and adaptation would have been critical if humanoids were to live to adapt at all. During this period, cannibalism may well have been necessary and would have played a part in the natural selection process. In the earliest days of life entirely underground, those traits which were best adapted to survival in the dark would have insured their perpetuation: A keen sense of smell, excellent hearing, quick reflexes and efficient metabolisms would have been perpetuating assets.

The terms troglobite and troglodyte mean distinctly different things; the former resides exclusively in caves while the latter uses caves as shelter but actually is more adapted to the outside world. Neanderthal Man, the best known and studied troglodyte fits the popular image we have of primitive man. We even associate the terms troglodyte and Neanderthal with a large, lumbering, coarse brute. However there is considerable recent evidence that these early cave dwellers were highly-intelligent, resourceful, thoughtful and artistic. They made clothing, tools, shelters and ritually buried their dead.

Even more advanced than Neanderthals, were the Cro-Magnon Homo sapiens which supplanted them in Europe around 32,000 to 33,000 years ago. Cro-Magnons had rudimentary forms of symbolic notation and even musical instruments. They developed rather refined methods of toolmaking, often utilizing materials gathered far distant from their homes. In addition to cave art, they ornamented their bodies and decorated other objects and utensils. They also hunted and fished in more refined ways than Neanderthals. In addition, there is evidence that they cared for and kept as part of their social structure, those too elderly or disabled to care for themselves. They styled their hair, had elaborate funeral rituals, probably had belief systems and may even have had a crude calendar. Many archaeologists and anthropologists think that these people were as intelligent or possibly even more intelligent than we.

Exactly how Cro-Magnon replaced Neanderthals is a topic of considerable debate among archaeologists; some think Neanderthals evolved into or were absorbed by Cro-Magnons, while others think the time span is too short for such a change and that Cro-Magnon migrated from the east and became the dominant population in Europe through superior adaptive abilities. In any case, the time for which evidence supports the simultaneous existence of both is very short - only a few thousand years. One interesting feature about the disappearance of Neanderthals is that very few actual bones remain (that have been found). Most of what is known about them and their lifestyle is through the interpretation of other physical relics: tools, weapons, structures, artwork, etc. Exactly what happened to them, why they disappeared is still a mystery.

Both peoples used caves however and interestingly, much of the artwork that has been discovered in more than 110 caves throughout Europe is situated in the deepest recesses, where there is absolutely no light. Torches had to be used to illuminate the painstaking work of creating the elaborate drawings that have been found on walls and ceilings. Why they choose to work under the most difficult of conditions is unknown. Was it that they felt safer, deeper inside the Earth? Did they feel closer to the nurturing Earth? Whatever the reason, they were drawn to do their most intimate and creative acts, not near the entrance of the caves, but deep within.

The apparent differences between known troglodytes and possible human troglobites are striking. Millennia of cave adaptation would have rendered the two even more far apart than we are from Neanderthal man. But these are the visible differences. There would likely be no loss of mental capacity. In fact, it is likely that the cunning and resourcefulness required to prevail underground may have resulted in intellect greater than ours.

Reduced caloric consumption and lowered metabolic activity would not have resulted in the loss of mental functions as it would diminished physical strength. Studies with animals, calorie-deprived and slowly starved to death, show that brain cells are the absolute last to be destroyed. Fat is consumed first, next muscle tissue, then other organs suffer, but brain function remains normal to very nearly the end of life.

So we would expect humanoid troglobites to be small with slender, hairless bodies and minimal physical development. Their large heads would be bald and have no distinguishing facial hair. Their very large eyes would probably appear completely black, and very large ears or hearing receptors would occupy a large expanse on the sides of the head and may or may not or be well-flared. Their voices would be high-pitched and may more resemble squeaks than speech. They would be highly intelligent and with a well-developed social order, for if they survived at all, social coherence would have been necessary. Limited space and resources would have dictated efficient utilization.

Would their world be populated by creatures besides themselves? Possibly, but not in the variety that we are accustomed. If humans or hominids migrated underground, it is quite possible, even likely, that other creatures did also. The same circumstances that allowed people to inhabit the underground would have also applied to smaller, less intelligent animals. However, from the cave environments we know and have studied, the underground food cycle is much more limited, and both numbers and types of flora and fauna are many fewer. In addition, during the early period of adaptation to the interior, humans may have depleted many of whatever other animals were trapped with them, causing the extinction of all but a few.

However, the entire hypothesis of a subterranean population depends on sources of energy and food which are unlike any found in caves that we know. If humans were to survive at all, sources of food from the ocean or from geothermal or chemosynthetic sources must exist. If that is the case, then other animals may also have survived. There may be a limited number of lower animals which are part of the food chain or may even be domesticated as pets or used as servants or assistants.

The idea of training animals to perform simple, routine tasks such as household duties, or to function as personal assistants or perform some lower operations in the professional world may seem strange because we have not fully exploited the potential of animals for these purposes. We have used animals almost solely as beasts of burden, making use only of their superior muscle power. However, the Romans trained apes to serve food tables and we have used pigeons as effective message carriers long before telegraphs and telephones were invented. The reliability and loyalty of dogs is exploited in their use as seeing-eye aids for the blind and their keen sense of smell is used in tracking and for bomb and drug detection.

Our use of animals has not been more extensive, in part because we have a long history of slavery which has provided ample supplies of humans to perform duties which higher animals might. Even in modern times, when slavery technically has been abolished in most civilized societies, there remain in effect social orders which effectively subjugate certain classes of individuals to the extent that there still is an abundance of people who can be trained to do such work. In a society in which every single individual must be developed as fully as possible for the functioning of a civilization with a limited population, the waste of human potential involved in allowing them to perform tasks in which animals can substitute, could not be permitted. Such an environment may well feature animals functioning at levels of competence higher than we are accustomed.

These would have to be animals of higher intelligence, of course. Perhaps there are subhuman primates which have experienced evolution along with the humans to become a subordinate, slave-class. Initially they may have even served as the hunter-gatherers, locating food which would then be forcibly shared with their human masters. A symbiotic operation similar to this is used to locate water in parts of Africa even today: An African bushman will tie-up a monkey until it is desperately thirsty, then unleash it and follow it as it locates a source of water. If underground humans used lower animals in such a way, it would have made them invaluable, at least initially, and helped both to insure their continued survival and establish their use as minions.

GEOTECH

If some specimens of primitive man survived beneath the Earth to evolve into a different humanoid species, it would be easy to assume that the darkness of the environment would have created such severe limitations that we might intuitively imagine them to be merely different looking descendants of primitive man, but "stuck in time" technologically, not much more advanced than they were at the time of their separation from our species. However, adaptation to their environment would have involved some technological innovation just as ours has. We cannot assume that these creatures would have remained primitive and helpless, clumsily surviving generation after generation in a dark, inhospitable world, unable to improve their lot.

Technology requires intelligence, physical ability to manipulate elements of the environment, the ability to communicate, written, symbolic forms of that communication, suitable raw materials and energy. We know without doubt that early man possessed considerable mental ability and had the ability to communicate (man alone can speak in the true sense, although the jury is still out on the question of the dolphins). It now appears that the ability to communicate verbally may be even older than previously thought. Until recently, the consensus seemed to be that speech developed in Homo sapiens with Cro-Magnon man about 30,000 years ago and that Neanderthal man was physiologically incapable of speech. The discovery, in 1989, in a cave in Israel, of the remains of Neanderthals which included the hyoid bone (the essential equipment for sophisticated verbal production) seems to indicate otherwise. If the implication is so, it means that speech has been a part of man's life for at least 60,000 years, probably much more. Prehensile hands are even older, so that even if separation from our species took place much earlier than the last ice age, our most ancient ancestors fully possessed all of the necessary physical and mental attributes to wrest forth a technology from the most barren conditions. But would the other essentials be available?

Raw materials are plentiful in the Earth's crust. Most of the elements which are so much a part of our modern world, come from the Earth itself. Iron, Aluminum, Silicon, Gold, Silver, Tin, Copper, Zinc, Lead, Carbon are all available in the Earth's crust in varying abundance. In fact, since water and ice cover more than 80% of the surface, restricting our access to the actual crust, many mineral elements may be more accessible from underground than from above. Even petroleum is probably more readily obtained from beneath than from above, though the application of fossil materials such a coal, oil and gas, if utilized by underground cultures, would necessarily be for a vastly different purpose, as I will explain later.

Energy is the most available resource needed for the development of an underground technological culture. Geothermal energy, pure heat directly from the mantle of the planet, is the simplest, most abundant, cleanest form of energy one can imagine, and absolutely free; but it is mostly unavailable to us on the surface (at least in most places and with our current technology). Geothermal energy, for all of these reasons, has long been the dream of futurists, but the cost of extracting it, at this point in time, makes its use entirely unfeasible in most places. However, to an underground culture, it may lay open and as ready to serve as fire, when first discovered by primitive man, and without the necessity of learning to initiate the process.

Geothermal power is completely clean in itself, which would be an absolute necessity in a confined world without the cleansing atmosphere that we enjoy and which permits combustion of fossil fuels with its concomitant release of Carbon Dioxide, Carbon Monoxide and other pollutants.

Geothermal energy directly in the form of heat may also be a solution to the Oxygen requirements of life underground. As mentioned before, underground spaces may have developed at the edges of the continental shelves, and may be partially inundated or be fed salt water through natural aquifers. It may be this interaction which fuels the creation of the atmosphere.

We are accustomed to associating technology with energy. That the development of technology must be associated with prodigious energy consumption and waste need not be so. We have enjoyed an abundance of readily consumable forms of energy from wood to coal to petroleum which has encouraged our present excessive appetite. Much of our use of energy, especially earlier in our history, has been for heat (later for light and cooling). A race of people without need for light or the need to use fuel for heating or cooling would have far less need for great sources of power than we do. If fact, burning of fossil fuels would be impossible; Oxygen requirements for combustion and the release of Carbon Dioxide, Carbon Monoxide and other pollutants would extinguish all animal life.

In a climatically hospitable and stable environment, the need for heat would be nonexistent. If underground creatures had vision sensitive to infrared or other radiation instead of visible light or if chemical forms of light developed that are similar to processes found in bioluminescent organisms, there would be no need for light to be derived from electricity. Electricity therefore, may well not have developed as a heavy energy source but might only be limited to low-current applications. It is possible that electricity from some form of chemical battery would have developed.

A device known as the Baghdad battery, a wet-cell chemical battery made of clay, Copper, Lead and asphalt, discovered in ruins dated between 248 and 226 BC in present-day Iraq, shows that artificially-created electricity existed far sooner than anyone previously thought and long predated any fossil fuel methods of generating electricity. Had we not possessed the capacity to generate large amounts of electricity through the burning of fossil fuels, we may have never utilized greater amounts than could be supplied by such chemical sources.

Many of our sophisticated instrumentation, communication and computational devices such as computers, are relatively low consumers of electrical current. Most existed in earlier versions which used far more electricity, the majority of which was wasted as heat given off by vacuum tubes. But the evolution of this kind of technology need not have been through a sequential chain of progress that involved high-energy prototypes. The transistor could just as well have been invented before the vacuum tube.

In some cases advanced calculations can be made even without electricity. The Antikythera Mechanism, an artifact recovered off the coast of Crete in 1900 contains a clockwork-like mechanism with very refined gearworks that seems to have been an astronomical calculator of some sort. It has now been authenticated as being from around the time of 80 BC Its exact use is conjecture but complex gearworks this advanced did not reappear until the middle of the 16th century in Europe. The abacus is another ancient manual ciphering aid capable of very involved calculation. With the application of only small amounts of electrical current and some expansion, it is but a few short steps away from a crude computer.

So much of our technological development has been marked by "bigness": We have enjoyed vast spaces and seemingly unlimited materials and energy and we have expanded our activities in every direction to utilize all that is available to us. Agriculture, food processing, construction, mining, manufacturing, retailing, transportation all require the movement of massive amounts of materials. As our world becomes more crowded, physical movement of people, materials, information, food - everything needed for life, will have to become more efficient and require less energy. We now physically transport virtually anything and everything over worldwide distances, not because it is absolutely necessary, but because the ability to do it is relatively cheap and we have had no incentive to work out more efficient systems. But already we are seeing diminution and energy saving design in an ever-increasing number of our tools. We are only at the beginning of this revolution.

During the 21st century, no doubt, communication will be almost exclusively electronic and the need for people to actually travel great distances in order to assemble will be drastically reduced. The physical transport of documents will be unnecessary for most purposes (the luxury we now enjoy of shipping massive amounts of paper, which ultimately occupies our landfills, is mostly for advertising - our mail system, unburdened of the responsibility for providing subsidized marketing services for business interests, could be freed to provide more essential services). Distribution of goods and services will become more regional, also minimizing the need to transport materials.

Throughout the past several centuries, the world over, we have expanded from local, to regional, to national, to global patterns of movement, with regard to the energy-consuming business of transporting people, food and materials. This will continue for some time, but inevitably it will be reversed. In the future, our connections to one another will be through communications and the movement over great distances, of anything not absolutely necessary to be moved, will be minimized. That we will have gone through such a period of energy wastefulness in our development will seem remarkable in retrospect. It has been so only because it was the easiest path to progress. A culture without such abundance could develop as well, only the means would be different.

In an underground world, with no vast open spaces, transportation over great distances might never have been a problem. All commerce and social activity might well have remained regional. Without great expanses of open land, there would no urgent need to go from place to place. Our needs for transportation are a result largely of the uneven distribution of resources caused by weather and the uniqueness of geographical locations. Perhaps within the confines of the inner world, life is more circumscribed.

Given the imagined conditions of subterranean environments, what specific technologies might we expect to have developed?

ENERGY

Advanced civilizations could exist without the abundance of energy that we enjoy. A technology which developed with far less than we have may actually be better off in the long run, for abundance invites excess, and many of our environmental predicaments are the results of energy abuses. But energy is not indispensable. Manufacturing or any conversion of materials from their natural states always requires some input of energy, no matter how efficient the processes.

At its most fundamental, energy is motion, either at the subatomic, atomic or molecular level, caused by one of the only four known (thus far) forces in the universe: gravity, electromagnetic, strong nuclear and weak nuclear, and in its purest form is manifested as heat. Other forms in which energy can be converted to do work, such as mechanical, electrical or chemical, always ultimately result from subatomic, atomic or molecular motion. Different manifestations of energy are convenient for various applications which require that the source be stable or portable or have other characteristics especially suited to the specific circumstances: Torches will work as portable sources of light but flashlights are better. A coal burning boiler with steam engine is a much more workable means of powering a ship or locomotive than an automobile, and would be even more impractical for directly powering a vacuum cleaner. The form energy takes is critical to the success of the device or process and in many ways governs the nature of technological advance.

Our earliest source of energy was fire, and in conjunction with water to create steam, powered our civilization's first engines. Steam, which was the backbone of the industrial revolution is still with us today for many industrial uses including the generation of electricity. With the discovery and development of more efficient fuels such as natural gas and various petroleum distillates, the internal combustion engine, either diesel of gasoline, made the power source more portable and allowed the development of the transportation-based economy the world now has. Without this compact, efficient power source, our economies, transportation patterns, and even cultures throughout the industrialized world would be significantly different.

Artificially generated electricity is relatively recent in our history. Although it appears to have been produced far earlier than previously thought (as shown by the Baghdad battery), the widespread practical application is hardly more than a hundred years old. Steam power was centralized and had to be transmitted to satellite users via mechanical means. This was a severe limitation on certain kinds of factories and manufacturing processes. Electricity greatly streamlined and expanded the range over which power could be distributed. The source of power is still centralized however, electricity is just more readily transmitted than mechanical energy.

An underground technology which had at its disposal, large amounts of geothermal heat, would probably have centralized sources of power as well. In a limited geographical area, this would be less a disadvantage than it is for us. A byproduct of energy transmission and use (whether electrical or mechanical) is heat, usually unwanted and wasted. The further from the source the energy must be sent, and the more of the transmission medium that is required (gears, belts, linkages, pipes or electrical wires), the greater the waste. Just as many of our early industries located on sites which provided the necessary energy for their operations (early mills situated beside streams for example), underground industries may be clustered around geothermal sources of heat with other energy needs being met by smaller independent or portable sources.

It's likely that such a civilization would have devised methods for utilizing steam as a direct source of power, or they too may use it to generate electricity. However their need for the flexibility of electricity may be minimal. Much of our use of electricity is for heating, cooling and lighting. Temperature control may not be needed and if light is provided by chemical reactions, the need for transmitting electricity is greatly reduced. Known chemical methods of generating light are more than 99% efficient - almost no energy is lost as heat. If underground people have retained some form of vision responsive to any part the visible spectrum, it is sure to be more sensitive than ours, and their need for artificial light could be met by such sources. Another factor in favor of the likely existence of chemical light is the fact that technology is often patterned after things observed in nature, either duplicating or extending them, and bioluminescence is a feature of many forms of life, both plant and animal, in environments with little or no light.

One can imagine a high-tech, low energy civilization which uses large amounts of power very efficiently and for only the most necessary purposes, while many other activities are carried out by relatively low energy methods. This is the direction in which our civilization may be finally moving and which ultimately may be partial salvation from the consequences of continued energy bingeing.

Undertakings for which large expenditures of energy are indispensable, such as changing the physical structure of the environment through excavation and the transport of large masses will always, and in any civilization, require the displacement of energy equal to the motion and mass involved. However, there is one other factor in the equation of energy required to do work, at least insofar as the concentrations of energy required - and that is *time*.

We are an impatient people, an impatience that has been supported and encouraged by easy and cheap oxidation reactions (combustible fuels and explosives) so we have not had to learn to wait. The amount of total energy required to do a job, say move mountain, is constant regardless of whether it is done in a year or in a millennium. However the size of the engine - the actual maximum output of the energy source required to do the job can be far less is the work is spread out over a longer period of time than if is has to be done quickly. A large enough explosion will move the mountain almost instantaneously, a thousand workmen with picks and shovels will take thousands of years, the erosion of wind and water may take billions of years. The amount of energy used is simply spread over longer time. No one can look at the Grand Canyon and not recognize the awesome total energy required to excavate that amount of material, however is was not accomplished at once.

Given the necessary mechanical leverage a one horsepower engine can move a locomotive - just not very fast. Of course as with all things, there is a point of diminishing returns. As more and more gears or pulleys or whatever means of converting motion is required to make a small power source do a huge job, the more is lost through friction and at some point would completely absorb the entire output of the engine. Still, the point is that much can be accomplished with far less energy than we are used to, if resistance can be minimized and one is willing - or must - wait.

It is this difficulty in recognizing the potential for low power sources that makes achievements such as the building of the pyramids by so-called "primitive" technologies so difficult for some to imagine; why mysterious intervention by an extraterrestrial entity must be involved. This is simply a bias resulting from our experience - the way we do things.

Just a the prisoner may dig a tunnel with a spoon, so might an underground people accomplish much with relatively small energy sources if they have learned to be persistent, patient and work together - and if they manage not to periodically destroy much of what they construct, necessitating equal or greater energy expenditures in rebuilding and replacement. This has been our way, but it may not be the only way.

If we consider how much of our constructive effort is continually being spent in rebuilding that which we destroy in wars, that which must be replaced because it was inadequately constructed in the first place, that which was damaged or destroyed by storms, floods, and fires, that which deteriorates because it was constructed from organic materials or that which we simply tire of because it no longer offers novelty, we can see that a great deal of our energy is consumed in redoing the same things over and again.

We have done it because our environment requires it and we because could afford the luxury. An underground environment would not have the effects of weather to shorten the life of structures, and surely building materials would be of earthen materials and not wood, reducing the loss by fire and rot. If they do not self-destruct much of their own creation with explosives introduced by warfare, they may only build once, drastically reducing their energy requirements.

In the long run does it take more to slowly construct durable structures only once or quickly build flimsy structures which must be demolished and rebuilt again and again? We think of our buildings as being almost as disposable as most of the other products we use, only the cycle of use is slightly longer. And since we have so few connections to our past, we think of building construction as something which must be completed in one generation. The concept of creating edifices which we will never see or enjoy ourselves is alien to us. Our cities are built over the course of hundreds of years, but these are not conceived as single projects, rather they are an aggregation of individual structures which arose in juxtaposition out of the needs of economic and social activity.

In a different social order, if the established pattern of building, imposed by limitations in energy, were spread over hundreds or thousands of years, the individual structures would likely be more valued and the generational bond established might make this an acceptable and satisfying way to create.

Transportation

Transportation on the surface land areas of the world is relatively straightforward. Since earliest times, man has walked or run, probably motivated by flight from predators and pursuit of prey. A life underground would likely have involved different patterns of physical activity.

Underground terrain, not being affected by the weathering influences of wind, rain, changes of seasons and so forth, would be generally more uneven than most areas of the surface. At least this is the case with most of the underground environments now known, and there is no reason to believe that others not yet located would be significantly different. In these spaces, humans would not be able to move as fast or freely as on the surface, and foot travel would more resemble climbing than walking. Obtaining food would probably be mostly a matter of foraging or gathering rather than hunting and pursuing game animals. Instead of large scale hunts, during which sizable beasts could be killed and shared with others, food gathering may have been a more consuming preoccupation that included all members of the population as soon as and for as long as they were able.

If areas of these habitats were filled with water which contained fish or edible marine life of any kind, swimming and diving may have been as much a part of their activities as walking and running has been for us. The early development of their technology to make food sources more abundant and easier to obtain might have centered around improvements in collecting, cultivating, harvesting, trapping and fishing instead of hunting. They may have developed some type of agrarian food orientation sooner than we.

Our technology, both with regard to food and transportation, has been completely influenced by our environment: Humans first rode animals to make their movements speedier and less tiresome. If sea creatures existed which remained entirely on the surface of the water and were large yet docile enough to allow them to be ridden, no doubt some aquatic form of animal riding would have predated the development of boats. Surely, early humans first learned to use primitive, natural boats by riding on logs and other buoyant materials found floating in waters. All man-powered boats after that were simply refinements.

For us, artificial means of rapid transportation over land first made use of the wheel - and still does. The observation of rocks or logs rolling down an incline may have been the source of recognition that round objects facilitate movement over flat surfaces. It is likely that these objects were utilized before the actual wheel was developed.

The progress in water transportation seems obvious, for the building of rafts, boats and canoes is easily traced from logs, dugouts and reed or straw bundles. Just as in the case of land travel, powered water travel first made use of animals' or man's own muscles. Then the energy of prevailing winds was used for centuries before the coming of steam, gasoline and electric power.

Air travel (space travel notwithstanding) has been the last barrier to our complete freedom of terrestrial movement. Though man has eternally envied the birds, movement through the air is technically the most difficult to achieve. And man-powered flight has only recently been accomplished and is still not practical. Animal power for artificial flight by man has not been done, though often a fantasy in the 18th and 19th centuries, demonstrating how extrapolation of a given application to different circumstances may not be workable. It is interesting to note that, here too, our earliest efforts at flight often attempted to emulate actions observed in nature, but in this case, all such efforts failed.

What sort of technical advances in transportation would be possible for an underground race? Paradoxically, it may have been that flight, of a sort, would be the first. In an underground environment, rapid motion over extremely uneven floor surfaces would not be practical with wheels of any sort. In fact, the kind of environmental observations that led to the discovery (or development?) of the rolling motion of the wheel may have been exceedingly rare. Stones which are quite round are mostly the product of surface environmental factors: weather, wind, water, etc. and logs would not exist. Large creatures which could be mounted, also would not be available underground. Humans would have been the largest animals such conditions would have permitted, and even they almost assuredly had to reduce their activity and calorie consumption in order to survive.

Birds as we know them, soaring as high as we can see, also would not have existed. However another phenomenon may have led to assisted ascent of some kind. If geothermal sources of heat were found, at some point the observation would have been made that materials thrown (or accidentally falling) over the rising heat would themselves, be lifted to the upper reaches of the area. Eventually, this could have been duplicated or harnessed to make floatation in air their earliest form of assisted transportation.

In a cave environment, weather in the form of disruptive storms and winds are not the hazard to lighter than air craft that they are in our open skies. In addition to the difficulties of powerless navigation in the atmosphere, a major problem of movement via floatation is the matter of limiting ascent and controlling the rate of descent. Underground, these also are much less factors. Ascent, which may only be for a few feet, or a few hundred at most, would be stopped by the upper limits of the space, and since the distance is short and the effects of atmospheric cooling are not encountered underground, descent would be slow and predictable. This would make hot air ballooning a gentle affair in which the rider might ambulate along, bouncing from the floor (or the ceiling) able to access locations for exploration or food that would otherwise be impossible to reach. Maneuverability in all directions would be much more an asset in negotiating irregular cavernous confines than sheer speed or altitude capability.

Another possibility for assisted underground travel is the large-scale creation of networks of scaffolds or "airwalks" which would connect different points, just as do our railroads, highways and sidewalks. Underground people would have even more need of specially constructed pathways then we. In our history of transportation, we have had enough extra land that before the development of high explosives and efficient earth-moving machines, we could simply build our roads around major surface obstacles. The terrain underground which is surely mostly uneven, would allow few options for avoidance. The irregular surfaces would require leveling or the creation of elevated roads.

Restrictions on available space would likely demand that structures be built onto and into the sides and even the ceilings of the caverns. Suspended pathways would easily connect different levels just as elevated corridors connect the upper stories of high-rise buildings in our major cities. In such a system, the roadway may be a stationary conduit for moving vehicles - trams or the like, or the roadways themselves may move, like conveyor belts, on which cargo, enclosed cabs and pedestrians might simply ride. We have used similar systems of cable tramways in extremely mountainous regions.

If direct geothermal power, or the geothermal generation of electricity is the primary source of energy for their world, the "conveyor belt" method of transport would be more efficient than individualized carriers. We have developed personal transportation because the luxury of abundant fossil fuels allowed it and the need to cross great distances made extensive common carrier networks impractical. Such would not be the case in their world. Furthermore, their culture would no doubt have developed social conventions consistent with the limitations of their environmental conditions and individual freedoms would be more limited along with the expectation of independent travel.

One can imagine networks of moving roads and sidewalks, crisscrossing to all levels of heavily populated larger openings, with single belts passing through restricted spaces or tunnels for longer distances, connecting major population centers just as rail lines and interstate highway systems connect our major cities. It would have been natural for them to expand and excavate, and many of their habitats and pathways may be artificially created.

Underground, just as on the surface, the physical properties of water dictate the nature of floating transportation. However, it is unlikely that bodies of water larger than small lakes exist in any openings under the Earth, and so long distance water travel, with the need to withstand rough seas, would likely not exist. Also, as with air travel, storms would not be a hazard to water travel, allowing the construction of relatively lightweight, fragile craft.

In order to fully explore their world, their need was probably to find openings to connecting spaces rather than to traverse the surfaces of great oceans. Since any connections from one space to another may be under water, submersible craft may have developed sooner than in our technology and may be the primary form of water transportation. Or as an alternative, there may be floating or elevated pathways over the water and tunnels which connect to other caverns through which conveyer belts move.

Cargo may be shipped via the conveyor belt system entirely with no need for discrete modes of transport. Such systems would work far better for limited distances, without the waste and congestion caused by traffic and competition for space. We have found such methods better in similar situations; for example the baggage delivery systems of large airports.

Any individual vehicles, whether air, land or water, would probably be quite small and light, possibly accommodating only a few creatures as passengers and with very limited cargo space. These may not be self-powered, but simply be coaches (for privacy in certain situations or for use by a select few important dignitaries) which ride on the conveyor belt system.

COMMUNICATIONS

What kind of communication would be possible in such an environment? Certainly radio and television waves would have limited usefulness. Our systems of global communications largely depend on the recent advance of geosynchronous satellite orbits which relay, around the curvature of the Earth, signals transmitted up to them from ground stations. Worldwide short-wave radio communication existed prior to this, since the range of frequencies used are reflected by both the ionosphere and the ground. In this way, despite the fact that they travel in a straight line, the waves are forced around the horizon by being bounced from ionosphere to the ground and back to ionosphere, back to the ground, and so forth. Irregular underground interiors would make this kind of containment propagation unreliable. Of course, the distances involved would be extremely short by comparison to our world. So some form of electromagnetic wave communication may exist.

Solid wire or fiber optic systems are completely possible and may well exist, however an even more straightforward medium could have evolved naturally. Solids and liquids such as rock or water are effective conductors of sound waves. Sound conduction is actually least effective in the atmosphere, which has been the main reason that electromagnetic waves and solid wire telecommunication has been so emphasized in our communications technology. Underground, a system of communication which depends on the encoding and decoding of mechanical transmission through rock or water could carry multiple messages through the same conducting surface.

Our current systems of telephone communications simultaneously send thousands of separate messages over the same wires or fibers and can be decoded back into individual messages. In a sense, radio and television likewise exist simultaneously in the same medium - space, (called the airwaves for purposed of legislative definition) yet are differentiated by the receiving stations.

The interior of the Earth itself could be used as a means of transmission, with multiple signals being transmitted through the same mass. The essential component for such a system to work would be a means of discriminating multiple signals. The ability to distinguish individual pitches which are being masked by an anomalous mass of mixed frequencies (called white noise) has been demonstrated. In fact, this is now thought to be a feature of the hearing and brain function of some animals, as a means of detecting danger amid the clutter of environmental sounds. Such a system of communication might use some combination of pitch, duration and sequential coding as discriminating elements in the encoding of messages. Range would be limited, but the system would require very little electrical current.

It would also have grown naturally from the most primitive type of communication - the beating on the stone walls of the environment. In our history, primitive peoples around the world have used the beating of logs, hollow trees, drums, etc. as a means of communicating through the air. Solid materials transmit sound waves even better than air and almost surely would have been used as a medium of communication, especially since the nature of underground space would have brought individuals out of visual contact with one another during exploration or foraging. Some spaces underground may be accessed only through underwater openings, and a communications system of this type would have allowed continuous contact while a diver was submerged as well as while they ventured into nearby spaces. Just as in the air though, the message is broadcast and can be received everywhere - the explorer had only to put an ear to a wall anywhere to intercept the message. In time, perhaps a hard wire system supplanted the walls of the caves for increased range and reliability. The actual messages may have remained coded or a transducer such is used on our telephones may have replaced codes with voice transmission.

The fact that we have detected no such sounds of underground life is not significant. For us to detect evidence of communication would require that we be listening AND that we knew exactly how and what to listen for. There is no shortage of sound being transmitted underground. The movement of magma, water, oil, plate shifts too small to be considered distinct events, in addition to the thousands of small earthquakes which occur worldwide each day, all contribute to a mix of underground noise that is analogous the radio signals which come from space. We know the significance of only a fragment of the interstellar noise and even less about the underground sounds.

The question of video communication is more speculative since the means of visual sensing permits a greater range of possible technologies. If an underground species possess some degree of bioluminescence, and some sort of emulative chemical light has been developed, then a hardwire system of video communication not too dissimilar from ours may exist, albeit not broadcast via microwaves. If their visual sense utilizes infrared wavelengths, a system of visual communication is still possible, but may not have developed at all.

Of course, no system of any kind, visual of aural, may exist. This topic of discussion is to examine the range of possibilities if a more or less parallel stream of technological development has taken place between ourselves and our underground relatives, different by virtue of the constraints imposed by their environments and their biological attributes. Improvements in transportation and communications are fundamental to the growth of any civilization.

Communication of course begins with language, at least spoken, and some kind of recorded language is necessary for the full benefits of idea transmission to impact a society. For while a spoken-only language affords interpersonal communication, lack of a recorded form limits language both in time and space. Recorded forms of a language expand its potential by not restricting the individuals or groups between which the information is shared to occupy the same place at the same time. Recorded forms permit language to travel, so to speak, to different places and times.

I use the term "recorded language" as opposed to "written language" purposely. We cannot assume that a people with different visual capacities from us would use writing as the primary means of recording their thoughts. In a world with little or no light, a tactile or audible means of transcribing language would be more the norm.

Braille developed quite late in our history (it is always the case that those in the minority are the last to have their needs met by society), and it mimicked the action of reading with our eyes. That is, it exists on a flat page and is scanned by moving the fingers over the page in much the way we scan the page with our eyes; the page is stationary and our eyes move to focus on different words or groups of words in sequence. A tactile form of writing that did not develop as a substitute for visual reading could take a different form. The message may be recorded on a linear format such as string, tape or wire, with the coded words as indentations or ridges in a sequentially coded pattern which is then fed through the hands, making the act of reading seem more like handling prayer beads, but at a faster rate.

Information magnetically recorded on audio or video tape is a non-tactile technology which nonetheless records information as a bit stream. The "Wall Street ticker tape" was an early method of communication which recorded and transmitted coded information about stock values to brokers. It was similar to our hypothetical system in that the information was read on a continuous paper tape as opposed sheets of paper. It preceded the invention of the FAX by many decades, and although it could not communicate complex data such as visual images, a similar method would be quite sufficient for language information.

Another possibility is that the information may be coded on a ribbon of some type in a form that can be seen and is scanned by visual receptors rather than felt. If some form of heat differentiation is used to imprint the ribbon instead of light, the ribbon may function more like movie film than paper, with the information encoded through chemical deposits on the ribbon which would heat at different rates. If the ribbon were heated or cooled the images would be visible in infrared relief, so to speak, which could be scanned if it were passed across the eyes. This might be moved directly across the eyes or be "viewed" through something akin to a movieola device used for film editing, in this case the light being replaced with a heat source.

In such a society, "books" might consist of coils of tape or cord instead of volumes of pages. Much of our information now housed in libraries is in the form of microfilm or microfiche and, increasingly, on CD ROM format.

A symbolic, coded form of spoken language may not exist for these creatures. If a sufficiently effective means of storing actual voice messages developed first, there is no need for a "shorthand" written version. Reading and writing print are separate skills from speaking and understanding the spoken word. We all learn to speak and understand first, then read and write later as a requirement of our society. Writing is, in and of itself, not necessary. If our current technology of audio recording had developed much earlier, the need for a written version of our language may have been far less, and a much less sophisticated form of symbolism, used only for signs and other distance communication may have been all that would have developed.

The colorful names of pubs and other establishments in England in past centuries, which made descriptive references to animals and other visible objects common to the area, were created because they could be represented on signs by pictures, recognizable to anyone, foreigner or illiterate, who could not read or write English. Even now, we see increasing use of international pictographs to accommodate speakers of different languages.

At some future point in our development, written language may be replaced by voice-recorded-only forms of communication. As computers become more capable of voice recognition and speech, and expand translation capabilities, actual voice communication will become as efficient to store and retrieve as written text.

There is some precedent for this type of communications evolution. Music, as an aural form of artistic expression, existed first as spontaneous creations, happening at a unique point in time and space, never to be exactly duplicated. Throughout several centuries, a complex system of graphic representation developed to allow the recreation of musical events in a different place at a different time. Music notation then, is a kind of "blueprint" for reconstructing the sound. With the advent of methods for recording sound directly, and even more recently the use of digital computer technology, music can be recorded or directly produced, edited and duplicated without the need for any intermediary written notation (or at least only for a different kind, which better suits the technology). For many kinds of music, written notation is completely bypassed. Most popular music now is recorded with no existing music, or only with the barest shorthand versions to aid the memory until the actual sound is recorded.

In a different world, the need for a coded form of speech may never have existed.

MANUFACTURING

Manufacturing exists to serve our needs and as expressions of our desires. The most fundamental needs which must be served (beside air, food and water) are clothing, shelter and security (these are in many ways related), transportation, communications and medical care. Everything we create for these purposes is through the application of intelligence with the appropriate tools and using suitable materials. Materials and tools are intimately related for tools must be fashioned to effectively manipulate existing materials and materials are created which in turn improve the tools. Rather like the chicken and the egg paradox, the question of which preceded the other is moot. Our most primitive tools were also our most available materials: stone, wood and bone. These were hard and durable, but could be modified from their natural shape by bending, breaking, grinding, and cutting. This was the beginning of our technology.

Perhaps man's first instruments were weapons, a simple club - stone or bone, which was useful in obtaining food as well as providing protection, both from animals and hostile humans. Much of the advances in manufacturing technology from earliest history up to the present day, are dedicated to improvements in weapons. Paradoxically, they are no longer needed to secure food, but this has not inhibited their production. Testimony that perhaps the primary reason for their existence in the first place was not the need for food.

Probably the first shelters to be used, other than naturally occurring ones like caves, were either simple lean-tos, fashioned from branches or leaves as an extension of the shelter provided by trees, or animal skins placed over the body, and perhaps later, over a frame of branches as an extension of the shelter afforded by animal skin coverings (whether early man huddled naked in simple structures or first used animal skin covering as portable shelter is not known).

What kind of naturally occurring materials would be available to an underground people for the growth of their manufacturing technology? It is unlikely that land animals larger than man could be supported underground. However, if underground spaces were connected to the open sea, large fish and even water mammals could be a part of the food chain and provide materials for clothing and tools. Over millennia, underground peoples' refinements in techniques and the development of new, artificial materials would have paralleled ours, for many of the elements available to us would have been available to them.

With energy and an abundance inorganic substances available to them, but with relatively fewer organic sources such as wood, plants and animal products, they may have developed synthetic materials like plastics or molded products much earlier than we. Soft stone can easily be ground to a powder which can be mixed with bonding agents and molded into complex shapes. Just as we use brick, adobe, clay and other earthenware materials, underground man may possess forms of molded materials as sophisticated as ours. In fact, our most widely used and durable building material, concrete, is produced from common substances in the Earth: sand, gravel and various forms of calcium derived from crushed clay and chalk. Concrete is not a recent invention. It was introduced to the world by the Romans and some of their structures have lasted for nearly 4,000 years. It may be the best all-around building technique yet known to man.

Metals are also a distinct possibility. Aluminum is the most common metallic element in the Earth's crust, consisting chiefly in deposits of bauxite ore. Although the current method of production involving electrolysis requires considerable electrical current, there are other methods which are now being researched, including a "carbothermic" process in which the oxide is heated with carbon with no direct electricity needed. A technology which had access to large amounts of geothermal energy could make use of a similar procedure for the production of aluminum, even if their capacity for generating electrical energy were meager. In fact, pure heat energy is the primary requirement for the production of most metals. Excellent bronze, iron and steel were made by the most primitive methods imaginable, before any means of energy other than burning wood or coal were available. With geothermal sources of intense heat, extremely high quality metals and alloys could be manufactured underground.

The mining of ores or other earthen materials might serve dual purposes; necessary construction materials are obtained, while at the same time additional living space is created. Just as we shape and transform our world, so would they. Our settlement of virgin lands often includes clearing forests which are then used in building structures. Historically, pioneer homes and furnishings were fashioned of whatever native materials were at hand; wood in many areas, stone, adobe, or thatch in others. Much of our modern technology still requires the removal of large amounts of earth and rock for use in building and manufacturing processes. Paradoxically, strip mining, which to us is a disfiguring procedure that leaves much of the land unusable, would be a functional process underground.

Oil, now so precious to us, may be completely useless in underground technology, at least as an energy source. It would not be surprising if it were utilized in other ways, however. Even our early civilizations used pitch and tar found in surface pools for paving material centuries before contemporary drilling for oil for use as a motor fuel. Underground people may have an excess which is drained to other areas, thereby reclaiming space needed for habitation.

The need for shelter underground would not be for protection from the elements. Temperature and humidity are remarkably stable, at least in the cave environments we have discovered. Depending on the proximity of the magma or geothermal water flows, some daily or seasonal variation in conditions may occur. But the most probable need for shelter would be for security and privacy, assuming that traits which make these necessary for us would also make them requirements in their society. Private dwelling spaces then, would not have to be constructed to withstand the force of weather or to insulate from the heat or cold. Security also may not require fortress-like buttresses since large animal predators would not exist. If social organization has minimized criminal behavior, living compartments may be simply carved out spaces in the Earth, with only minimal artificial facades erected for enclosure. The wood on which we have depended so heavily for home construction would not be available nor would it be needed in this kind of world. Many primitive cultures live successfully without massive structures to protect them from one another.

Clothing also would not have to protect individuals from the elements and could serve strictly as social convention and personal decoration. Many unmodernized peoples which populate temperate areas of our world wear little clothing. It may be that, in a culture in which waste is not a luxury, human skin or bones may be an accepted part of the attire. The weaving of hair or other fibers or even the creation of synthetic fabrics is entirely possible.

Textiles woven from crops such as cotton obviously would not exist, but other flexible materials may be made into crude cloth. A variety of waste materials might be used in making paper or similar materials. If areas of their habitats are flooded by the sea, many by-products of marine life could be used, such as bones, scales and skins of fish. One of our most prized fabrics, silk, is after all the by-product of moth larvae.

Until very recently, almost all of our manufacturing processes made use of raw materials obtained from the natural environment. This happened for two reasons: One, because naturally-occurring materials far outnumbered manmade materials, and secondly, once processes were established for the extraction and conversion of virgin raw materials, it war far easier and cheaper to rely on these than on collection and remanufacturing of man-made objects. This is now beginning to change: As original sources of minerals and other natural raw materials becomes more scarce, as population growth makes the availability of manufactured materials more abundant, and as waste disposal becomes more difficult and expensive, it will become increasingly practical for industries to develop the technology needed to reuse previously manufactured objects.

We are only now at the beginning of the recycling era, and just as past historical epochs have been characterized as: stone age, bronze age, steel age, machine age, industrial revolution, or information age, the next era may become known as the "recycled age" or conservation age"; an age in which we make new and extended use of all that previous generations have created.

But this may have been going on for a long time underground. The dynamics of change are governed by the unique conditions of every environment. Underground, with a limited population and limited space, the ratio of easily-obtainable raw materials to the accumulation of artificially-created objects may have reached the balance point needed to force recycling long ago.

In a strange kind of irony, we accept (although not without resentment) that our own bodies are temporally finite. From an early age, we consciously and subconsciously plan our lives around an assumed life expectancy that seems the average. Our periods of education, entry into the productive workforce, marriage, procreation and retirement are all timed with the knowledge of approximately when we can expect to decline in powers and die (barring some unforeseen accident). And in religious rites which reflect our agricultural and predatory heritage, we symbolically recognize that the materials of our physical bodies may be used again in the natural biological cycle of birth and death. Increasingly, we are using the "spare parts", of our dead bodies; those organs which are still healthy and have outlasted disease or escaped injury, are sewn into living individuals whose corresponding components have failed. Our attitudes about this are changing from revulsion to acceptance to expectation.

However we have traditionally thought of manmade objects as junk, once they are outmoded or no longer functioning; a curious relic perhaps, a keepsake or an interesting antique, but seldom as part of a natural cycle, in which the same chemical ingredients are used over and over again in the same way that decaying matter provides food for life in the biological realm. As our technology becomes more sophisticated in this regard though, we will come to expect that all things, once made, will simply be converted into something else after their usefulness is past. All objects will be considered transitional, as indeed they are in the larger scheme of things.

Underground, manufacturing technology would have had to confront this reality earlier than we have. They would have little use for landfills. Waste disposal would be a critical matter for them. With no extra space, any used by leftover junk would occupy precious living space badly needed by a contained population. Their manufacturing process may already be based on re-use, and few if any artifacts remain of their past history. Archaeology as we know it may be unknown to them, and unless written records better than we have kept are a part of their tradition, they may have little knowledge of their past. Just as the biological food webs of cave environments are much more efficient than surface environments, so would the manufacturing systems also have to be.

AGRICULTURE

That the maintenance of adequate stores of food has occupied much of our creative energies is a reflection of the vagaries of our weather and climate-based food resources. Food crops are seasonal in nature, and each has its narrow specific parameters for successful cultivation. Added to this, the fact that many areas otherwise habitable, are unsuitable for any sort of agriculture, and those which are face the competition of other plant life and the ravages of insects, makes the business of securing food precarious indeed. For this reason, methods of preserving foodstuffs so that they will survive wide distribution and storage through lean periods has been, historically, a major human concern.

Before man learned to tend crops and heard animals, it was necessary to forage and hunt, migrating as necessary to more bountiful locations. This was a significant constraint on other aspects of social development, for thought, intellectual development and progress requires some idle time. If fresh food were to be available everywhere at all times, there would be no need at all for preservation techniques. Cooking animal flesh extended its "shelf life" slightly but the real freedom from the constant animal need to pursue food was the development of crop cultivation and the husbanding of heard animals. This allowed humans to remain in one location and stabilize their habits, building more permanent shelters and fixtures to enhance their lives, and give attention to other matters which needed improvement.

Various natural processes which transform perishable foodstuffs, through harmless microorganisms, into more stable, lasting edibles, were undoubtedly discovered by accident. Various pickling and fermentation processes are responsible for cheese, buttermilk, sauerkraut, yogurt, vinegar as well as alcoholic drinks. Drying, salting or packing with other spices also staved off dangerous bacterial spoilage in areas where ice was not available and before the development of artificial means of refrigeration. Canning methods are quite recent in our history and frozen foods date only from the 1950's. More recent still, are methods of irradiation and high temperature pasteurization to preserve foods. These all have been necessitated by the fact that, despite our many technological wonders, the food supply is still uneven and insecure.

An underground habitat, though lacking the nurturing energy of the sun, may actually be more conducive to regular food production than the surface. The sun, which fuels plant growth, is indirectly an agent in the extreme fluctuations of weather that so often threaten crop productivity. It is the uneven heating of various Earth surfaces, their differing densities, the curvature of the Earth and the angle and relative distance from the sun that contributes to the irregular supply of rainfall, storms, and daily and seasonal variations in growing conditions (for the elliptical nature of the Earth's orbit around the sun and the axial tilt are contributors to the seasons only because they change Earth's distance from the sun). Whatever growth that can take place in the absence of sunlight, would at least be uniform and non-seasonal.

Although crops of the kind we are accustomed to would not be possible, this does not mean that cultivation of food could not exist underground. If caverns are fed with supplies of seawater, aquaculture could have developed in place of soil based crops. This could take the form of animal protein growth such as fish farms or mollusk beds, or directly as cultivation of nonphotosynthesizing vegetable life. Any ocean waters would surely carry organic material and possible small fish into the habitats, and tidal movements may periodically replenish supplies of sea life to these underground locations. Although large fish may not pass through whatever openings allow the water to flow in and out of underground chambers, very small fish may, and may even migrate there as a refuge from larger predatory fish. If tidal movements provide circulating flows of seawater in and out of caverns, a form of fish harvesting may have developed which traps sea animals brought in at high tide by restricting their escape with the receding waters by lowering nets over the drainage orifices at the appropriate time in the tidal cycle.

Also possible is the cultivation, either in water or in particulate material, of various forms of Fungi (which lack Chlorophyll needed for photosynthesis), of which mushrooms and truffles may be the best-known examples. Cultivation of such crops would be independent of the sun and food production could be rather continuous. If organic material from the sea is introduced by cyclical flooding, crops of this kind could have evolved.

Of course there needs to be some form of organic nutrition for any kind of plant or animal life to flourish. The usual kind which enriches the soul on the surface; animal droppings, decaying plant and animal structures, would not be available underground. But in addition to organic material which might be brought in from the seas, another source might exist: As already mentioned, petroleum stores are readily available underground and are really nothing more than enormous mulch heaps of extremely advanced age. The fact that they are fuel for combustion is evidence that they can support life also, for combustion is simply a vastly accelerated version of the same chemical reaction which takes place within the bodies of animals. The fire of life is really a slow, controlled form of oxidation. The analogy might be drawn between a nuclear explosion and the controlled fission of nuclear reactors; both are the same process, only the latter is controlled while the former is uncontrolled. In chemical oxidation, if unlimited oxygen and pure fuel supplies are brought together in the right mixture, explosions result. Within organisms, metabolic reactions are limited (just as nuclear power reactors moderate fission) so only the mild heat of normal body temperatures are generated. An underground species, living in an environment with less oxygen than in our atmosphere, metabolic processes may be even more subdued and may feature lower body temperatures than most surface animals.

Remarkably, petroleum can be used not only as a source of energy for microorganism which can then support higher animal life - it can be used as a direct source of food for large animals and even humans! In a process that uses bacteria and yeasts, proteins, carbohydrates and fats can be synthesized directly from crude oil. This is not an exotic procedure requiring sophisticated modern techniques which can only take place experimentally in a laboratory, producing only minute amounts, but a relatively simple, natural process that produces protein at a cheaper cost than similar protein derived from skim milk. Petroleum-derived protein in a powdered form known as petroleum flour has been used to feed livestock in the former Soviet Union since the 1960's.

If the prospect of surviving on some kind of seaweed, microorganism, or oil-derived food seems unappetizing, we need only to remember the range of foodstuffs that are eaten by various cultures, indeed that are often considered delicacies, to remind us that the definition of acceptable food is an entirely cultural trait. An underground race would have evolved using whatever was available and to them this would seem normal.

We take the surface world to be nourishing and hospitable because, having evolved here, it seems natural to us. But in objective terms, it may be considered far less than ideal - and life has flourished in great variety, in spite of, not because of the uneven conditions. It my well be that in many ways, an underground environment is a more stable and nurturing place for some forms of life than the exterior world. At least food supplies there may be more constant, minimizing the need for preservation.

Let's consider some of barriers to the perpetuation of life on the surface which we seldom think about:

First, as already mentioned, only about 15-20 percent of the surface in habitable by land creatures without extraordinary measures which are possible only through modern technology. Extremes of climate and weather which make food supplies uncertain are responsible for floods, storms, famine and fires which have killed untold numbers of humans and animals throughout history.

The balance of interdependency between various species is extremely delicate. One would expect that the competitiveness for limited resources would ultimately lead to the absolute dominance of a superior species which would consume or destroy all inferior forms of life, and in turn would perish or self-destruct through habit of avarice. It now appears that human beings are capable of this, and it remains to be seen if we are actually in the process of doing so.

Ultraviolet (and other shorter wavelengths, such as cosmic and gamma radiation) is detrimental to cell health in all animals. Were it not for the combined protection of the ionization layers of the atmosphere and the protective coverings of all animals: fur, feathers, scales and melanin (in the case of humans), the radiation from the sun would make quite life impossible.

Yet despite all these barriers to survival, life has adapted and prevailed in some form or other. Fecundity seems to be extraordinarily powerful in overcoming adversity, and while the foodstuffs that may support life underground would seem unusual to us, nonetheless would provide the sustenance necessary for life to exist.

MEDICINE

Despite so many remarkable advances which have dramatically improved our lives, none is more welcome than effective medical treatments. Two of the advances which have most extended life expectancy and relieved suffering are quite recent: Antibiotics and anesthesia have diminished the horrors of bacterial infections and injuries since their introduction in the mid-19th century. Yet treatment and prevention of disease remain major concerns, for there still seems to be no end of debilitating malfunctions which can afflict us. Humans are not unique of course, all living things, plant and animal, suffer disruptions of normal functioning. These fall into three broad categories: damage brought about by injury, infectious diseases caused by microorganisms and genetic or other congenital flaws.

Throughout most of our history, injuries have been thought of as sudden traumas. Increasingly however, we are discovering that subtle, long-term injury may not manifest itself for years, decades or even within a single generation. Environmental factors, either natural of manmade, profoundly affect health and longevity and the resistance to infectious, and congenital diseases as well. They might be thought of as long, drawn-out injuries, from which the body's self-healing mechanisms never fully recover before additional injury is inflicted. Even aging, always considered to be unavoidable, is in many ways a product of, or at least is accelerated by long, slow damage caused by environmental factors and even our own metabolic processes.

In at least one important respect, underground life would be spared the continual damaging effects of our natural environment - most importantly the sun and to much lesser degrees the extremes of temperature and humidity. Paradoxically, the sun, which is the direct source of life on the surface, is very damaging to animal life. Is has seemed natural to us to worship its light (and warmth in cold climates). It obviously contributes to plant growth, so we historically have assumed that is good for us. We even use the term "healthy glow" to describe the skin's protective melanin response to solar exposure. This idea is slow to be replaced by the biological truth about the sun's radiant gifts; that the heat and light come at a high price for our health.

That the sun darkens the skin has long been known, but its contribution to drying, wrinkling and other physical deterioration has only been fully appreciated in the last few decades. The effects are not only cosmetic though. The sun's longer wavelength ultraviolets penetrate deep within tissues and diminish immune system functioning. The darkening of the skin after exposure is an adaptive response serving to protect underlying tissue from damage, but is not entirely effective. Long-term exposure to the sun compromises the optimal functioning of cells and makes the human organism more receptive to diseases of the skin and other organs.

Most studies which track solar exposure and relate it to the incidence of disease have been directed at malfunctions of the skin, but it's contribution to other diseases is theoretically established. It is likely that disorders previously not thought to be affected by outdoor solar exposure will also be found to be related. Underground creatures not be burdened by the stress of solar exposure and presumably would not have elevated incidence of diseases associated with the sun.

The amount of illness precipitated by other forms of natural and man-made environmental stress is incalculable. Microorganisms cause disease, but immune system response, the body's natural defenses against such invasion, is directly affected by extremes of temperature, both hot and cold as well as by physical and mental stress. Industrialization, particularly in the last century has contributed untold chemical pollutants to the environment.

These are not only absorbed by contact and breathed, but have now entered the food chain and are eaten and drunk as well.

In addition to not having to withstand the stresses of temperature extremes and other natural environmental factors, the inability to burn fossil fuels underground would also reduce inhaled pollutants. Since the cultivation of crops as we practice agriculture would be impossible, it is likely that the diet of an underground race would be almost completely lacking in saturated fats with the commensurate radical lowering of cardiovascular disease. If their cultivation of food sources is either of aquatic plant or animal life, their diets would be vastly healthier than ours and aquaculture requires no pesticides or herbicides .

As regards injuries, these too may be less frequent underground. If we consider how many accidents are directly or indirectly caused by the weather, or during farming or construction, or in ocean, land or air transportation, we can see that if a habitat restricts the possibilities for such activities, the rate of accidents would also be reduced. Our surface environment has a large scale - ambitious activities pit us against vast distances, powerful forces and massive objects. We are an energetic people, fueled by the opportunity that almost limitless area and materials afford. Our habits are slowly changing though, as we become more populated and the open areas which once permitted and invited unrestrained personal behavior are filling up with neighbors who place constraints on our activities. No Texas rancher feels the same confinement of his personal habits that a Manhattan apartment-dweller learns to live with. Unavoidably, close quarters restrict personal freedom, and it is a natural function of government, as populations become more dense, to respond to various constituencies by inhibiting the freedoms of others. The confinements of underground spaces may have long required minimization of physical activity and limitation of individual rights.

This is not to suggest that accident or disease would not exist. But in a society in which disease and injury are less frequent occurrences, treatment may be much less a priority. It is conceivable that no medical treatments exist at all. Some members of religious cults interpret medical intervention as an interruption of a natural process and refuse to be attended by doctors.

Self-concern usually preempts the acceptance of such philosophies. However, if population management is a major priority in a society which functions with limited resources, non-treatment may have evolved as a societally-reinforced measure. The sick are simply allowed to die. Treatment of infectious diseases which threaten the population as a whole may be dealt with through compulsory quarantine. Individual illness may be considered unfortunate, but may be as accepted a part of the vagaries of circumstance as poverty and financial ruin are considered in our culture. The absolute non-treatment of injury or disease is cruel by our standards, but if conditions have imposed stern measures for the ultimate survival of the whole, philosophies would have developed which glorified it just as economic philosophies justify monetary predation in our society.

Limitation of population size would undoubtedly be a primary concern and restrictions on births alone may not be sufficient to control growth. Some form of ritual eradication of sickly, injured or handicapped members of the society may even exist, if it were more societally functional than developing treatments to rehabilitate them.

ART, CULTURE, RECREATION AND ENTERTAINMENT

The inclusion of these topics in a single discussion is somewhat reflexive of convention, for they are often lumped together as though they are more similar than they are. They are really quite different things, but because the distinctions are difficult to define, and there is much overlapping, they are customarily discussed together. They do share certain similarities however, most fundamentally that these activities are expressions of impulses which have no other outlet and often are pursued entirely for the pleasure of doing them and not for practical gain. It is this impracticality that often causes their value to be contentious.

The role of these kinds of pursuits has varied considerably throughout history and in different cultures, and is variously regarded by different individuals now. Many people consider ballet, poetry, opera or classical music to be useless, pretentious frills, without which society could well function. Others may think that football, soccer, auto racing, golf, chess or basketball are complete wastes of time. Still others view watching movies and TV as escapes from reality - fantasy not worthy of responsible adults. That one person's passion is another person's trivial pastime is a measure of the egocentric perspectives of most of us. All of these activities are equally useless from the practical standpoint. None directly provides air, food, water, shelter, clothing, medical care or any of the goods and services needed for sustaining life. However, for the enjoyment of life they are indispensable. Their value lay in the exercise of mental and physical abilities which may not be called upon in the absolutely necessary operations of life that do provide the basics. This is their value to those who actively participate, and to those who watch or listen, any and all are forms of interested relaxation - entertainment.

Their existence serves to stimulate responses within us which cannot better be stimulated through other activities, and in that sense they expand our capabilities. This is one of the essential ingredients for the advancement of humanity; that we successfully learn to provide for our needs at one level, are then freed to devote more of our time to other pursuits and in so doing transform the world to another level of existence. Principles of law, government, economics, science, medicine, philosophy and virtually every other body of thought could not have developed until basic needs of life were met to a degree that allowed sufficient leisure for idle thought. It is the degree to which a society has fulfilled its most basic operations, permitting a greater portion of time by more of its members to be devoted these nonessential activities, that some consider to be a measure of the success or sophistication of a society.

It is the sum of these activities and the manner in which they manifest themselves that constitute a great part of what is commonly described as culture. Of course these are not the only measures by any means, for culture manifests itself in virtually every human endeavor, from religious beliefs to political organizations to product design. But as a convenient means of discussion, is interesting to focus on how an underground environment might influence the art, culture, recreation and entertainment of its inhabitants.

Let's start with the easy ones, like literature. We can, with reasonable confidence, assume that some form of literature will exist in the culture of any intelligent species capable of communication and symbolic representation of thoughts. All known civilizations have literary traditions that were borne out of the need, not only for self-expression - to tell a story, but also to pass on information - to tell others what happened, and in that sense, early literature was inseparable from history, and to some extent still provides insights into the historical periods during which material was written. So we can assume that underground people will have some form of written or recorded language and that stories will vary in merit, based on the talent of the storyteller. It is this distinction of excellence that is part of the criteria for literature.

Our literature consists of written records documenting essentially private personal commentaries. There are jointly authored books, of course, but even these are written with but a single voice. For multiple voices to appear, each with distinct and definable characters, but in an objective rather than subjective context, we turn to drama. However other forms of literary expression are possible: Narrative writing could be a collective process in which conversation and ideas are presented as discourse among several people with no dominant writer's voice. This form of writing is usually used for interviews and journalistic reporting in our culture, but it is conceivable that a hybrid form, something between a play and a novel exists, in which all characters contribute subjective narrative.

Another possibility is that literature could be interactive, just as video games and sports are. The reader (participant?) could interact with the story, altering it with their responses, making the act of reading a collaboration with the author(s), in effect making the readers part authors. Reading (or participating) in the story would become part of a continual creation in which the outcome was affected by those who had partaken of the experience, leaving their lasting mark on the constantly-changing product. About the only thing we have in contemporary society that approaches this additive process is graffiti. Serious literary works are highly egocentric and protected from alteration. In a different society, one in which the primacy of the individual is unknown, collectivism may be a feature of many art forms.

We have artistic endeavors which meld the talents of many into a single outcome: symphonic orchestras and theatrical movies are two examples. However even these which may include the talents of hundreds (thousands, in the case of films), the vision is clearly from a single perspective, and all participating individuals contribute only that which supports the single view. There is a paradox here; our society which gives considerable authority to the individual, requires sublimation in joint ventures, while a culture which is more collective and mutual in most of its operations, might share equally in joint projects without the need for a central controlling authority, liberating the act of creation.

As has already been mentioned in connection with communication, the physical form and organization of symbols used for transcribing may be quite different from our own (in fact, probably will be).

If recording of verbal/literary creations is by direct acoustic means without intermediary graphic representation, then collective discussions - improvisations of a sort, falling somewhere in between parlor games, theater and debates, with broad rules governing their progress, might exist as a consumable form of literature. We have a musical art form - jazz, which operates exactly this way. Such an arrangement for literary creations may function on many levels: as literature, entertainment, public commentary, news and personal emotional therapy. While this may seem strange, certain fairly new forms of entertainment within our own culture are functionally similar: audience participation/call-in radio and TV talk shows, while sensationalist in nature, nevertheless do represent a sampling of the attitudes and values of those participating. They apparently function as emotional and intellectual outlets for many who seem to have few other means of expression open to them. Furthermore, they are useful indicators of the attitudes of at least that segment of the population which takes part and may resonate through a larger audience. And although the topics discussed are chosen to be as provocative as possible, and little of enduring value is ever created in the melee, one can imagine that if planned and coordinated to a higher level of discourse, much useful thought might be distilled.

One arena, controlled by strict rules of presentation, on which we rely to forge meaningful conclusions and establish lasting precedent from collective efforts, is the trial procedure of our legal system. In another kind of world, legal matters might be decided be a single thinker, one whose authority is unquestionable while literary output could be the product of systematic, controlled, collective interactions.

Music presents many possibilities, much depending on the hearing of subterranean people. We can assume that they would have some form of hearing and the pitches in which most of their music lies would be driven by that. However it's possible that sensory perception of music might involve some direct physical contact with the instrument or vibrating body (for the audience as well as the performer). In this way the experience might be more intimately shared with the performer(s). Music might have evolved as inseparable from dance or some other kinetic expression of enjoyment. In this regard we can also assume that music would contain a rhythmic component. If fact it may contain *only* a rhythmic element. Melody, harmony and timbral variety are certainly important contributors to a full musical experience, however rhythm alone can be complex and satisfying. Various national forms of folk music and even contemporary folk music have little of the harmonic richness, melodic interest or timbral variety of cultivated concert music, yet seem to provide completely satisfying experiences for many.

We are accustomed to music being a product of ensembles for the most part. But in a different world, the uninterrupted playing of a single individual either improvising or performing planned works might be considered the highest form of musical art. If the subterranean culture is especially regimented and communal as suggested, then the opportunity to play music, unfettered by convention or the requirement to cooperate with others may be the ultimate personal freedom.

As far as what kind of instruments might be expected, vocal music of some kind would likely be one of the earliest and probably remaining genres, as well as music using sounds made with objects and materials common to their environment. It has been for nearly every known culture, and only quite recently has sound been synthesized from non-acoustic devices. Although the kinds of materials from which we have fashioned primitive musical instruments: wood, animal skin, animal horns, may not be available, other natural materials may: Stones, stalactites or stalagmites, or other resonant structures, even water, may have formed the instrumental basis of their musical heritage. Other body noises, grunts, growls, screams, slapping, stamping and so on, may be part of a music and dance in which the sounds of the act of dancing are part of the music, as it is in tap dance and some folk dances. In their culture, music could be a byproduct of dance and not the other way around as it is with us.

Music serves many functions in our modern society, ranging from elaborate presentations, which are given undivided attention, to popular music, listened to obliquely on car, office and home radios to "elevator music" which is not really intended to be listened to at all, but simply serves as atmosphere in business offices and restaurants. It's possible that this range would not exist within a different culture. Music may be thought of as a rather continuous contributor to the decor or ambiance of an environment, in which case the aspect of form, so important to art music which is created as a aural edifice - a single creation with a beginning, middle and conclusion, would be useless. Or the reverse could be the case, music could be considered so rare and special an event that it holds almost religious significance, to be attended with great ceremony and respect. After all, it has been the advent of recording, with the abundance of available music at any place at any time, that has, in part at least, led to its relegation to sonic decor status in many circumstances.

One of the major distinctions between arts such as literature, drama, music and dance, and those such as painting and sculpture, is that the latter exist in a static state while the former take place over time; they are temporal arts which rely on anticipation and memory for the sense of form they create. Various written symbols can serve to preserve the instructions for recreating the event and newer technologies such as film, audio and video recording and digital computer technologies can preserve the actual event. But painting and sculpture are lasting records in themselves. They require no further technology (except perhaps to prevent decay and deterioration) for them to remain a lasting part of the cultural environment. In this sense, both are more akin to building and architecture than are the other arts. For this reason, much of what we understand about vanished cultures is learned from their physical artifacts.

Sculpture for its own sake is closely related to more applied arts involved in the design of functional, everyday items. Painting too shares a functional aspect, for before video cameras and photographic processes, painting was the only means of documenting visual images. In more primitive societies, the functional aspects of these arts were indistinguishable from the aesthetic. As technology progressed and other methods supplanted these arts for functional purposes, they grew more abstract and were liberated to become pure expressions of thought and feeling, with no need to be representational. If we assume that this same dynamic applied to the evolution of art within an underground culture, we cannot know much about their fine arts without knowing the level of their technology. However, if we assume that their technology is sufficiently advanced to include powered craft (as will be discussed later) then it would follow that visual arts would have developed to a highly sophisticated stage involving some degree of abstraction.

Complicating the matter when considering visual arts is the unknown factor of vision and lighting. If the underground environment is absolutely dark and underground people have no visual capacity, then obviously no painting or other visual forms would exist and sculpture or some other kind of tactile art would be supreme. However if chemical, or even electrical sources of light were developed before visual senses completely vanished through disuse, then visual art which rely of luminescent rather than reflected light might exist: Various pigments which radiate different colors could make paintings "glow in the dark" images, and viewing them would be more like going to a movie or a slide show than visiting a gallery.

Any of the arts might have achieved an extra dimension of social importance by being allied with religion or medicine. Artists may be considered something like priests or shaman, with supernatural significance associated with artistic vision. It is thought that the early cave drawings of Cro-Magnon people arose out of a need to visualize successful hunting expeditions (if this is the case, then they predated the visualization techniques used by contemporary sports psychologists), give thanks to the gods, or for some other superstitious purpose. Visual arts have long been a vital part of religious tradition and even today musical expressions of holy tenets are a integral part of many religious rites. It may be that these associations are even more intimate in the practices of subterraneans.

Physical outlets and expressions are no less important than mental ones, and there are many areas in which these seemingly different kinds of activities share similar features. Play and games have always been a part of every culture and this need seems to be a trait that is not only human, but one that is held by many other animals as well. There is almost always some competitive element to games and sporting activities and this not surprising, since in the primitive sense, this kind of play serves to develop skills which directly bear on survival: Hunting, combat, pursuit, escape, endurance, agility, concentration, heightened awareness, are all mental states and physical aptitudes that have practical uses in more primitive cultures. Like art, these were once more closely associated with practical everyday actions than is now the case. In a sense, sporting and recreational activities are vestigial; they keep active, abilities which have little actual use in contemporary society. However, social development often precedes biological adaptation and so in many ways we remain biologically better suited to a less civilized and more active circumstance than many of us actually experience. In this respect, sports and athletic recreation serve a vital function: to keep our bodies in tune and provide an escape from stresses which have no outlets in civilized society.

To what degree these sorts of physical outlets exist in a subterranean people is difficult to imagine. If, as we have assumed, caloric intake and output is far more limited than in our species, activities involving the degree of strength and speed that we are accustomed to, likely would not exist. Their sports may be oriented more towards agility or fast reflexes. And If we further assume that their sports evolved from activities which had their origins in useful primitive skills, as has ours, then we must consider what those skills would have been in such an environment.

Certainly good hearing and the ability to identify and accurately locate sounds would be an asset, so some form of game which involves finding an object by sound clues only might exist. A keen sense of smell would also be valuable (as it is in surface animals) and games in which something is identified and located only through scent - a kind of tracking/scavenger hunt competition might also be one of their events. The ability to navigate in the dark, through a complex maze of underground tunnels and remember pathways taken could be the difference between survival and death, so some kinds of navigating competitions could have survived into contemporary sporting events.

If we assume that in order to have survived this long, theirs would be a homogeneous culture with relatively few internal conflicts. And that in a world with limited space they faced fewer threats from adversarial civilizations, then it is unlikely that sports which feature any kind of physical combat exist. Running is related to conflict and is also impractical in cave environments and so team sports which feature running are also unlikely.

Water sports are a good possibility, if as we have assumed, the environment features large areas of water. These may involve competition over depths attained or ability to remained submerged for long periods (both useful skills in exploring and gathering marine life). Water sports might also include some aspect of location or navigation.

These and other sports may, as in our cultures, be vestiges of the past which are physiologically useful but which also serve to maintain emotional bonds with that which once was.

HEAVEN IN EARTH

If we assume that an underground people would share the same basic genetic background as we humans, we must also assume that some religious or philosophical beliefs would have developed. It may not be that they are currently practiced, but they would have appeared at least at some stage of their growth, for religious beliefs play a vastly different role in the lives of different individuals and cultures at various times. In most, beliefs are rigidly fixed. In fact, most religions require this as a condition of participation and acceptance, in addition to the adherence to other tenets of the faith. Temporary questioning may be allowed but, ultimately, few religions will permit the expression of long-term, avowed, ambivalent attitudes. In other individuals or peoples, or at different times in course of their experiences, religion occupies a less prominent position in their conscious concerns.

Certain fundamental perspectives as well as practical elements in the observation of the faith are influenced by physical environment. This is the case with our religions and it must be assumed so for any earthly people, terrestrial or subterrestrial.

So what might the effect be of an underground environment on the presumed beliefs of an underground people? We can safely assume that the concept of heavenly deities, gods, angels and the like which figure so prominently in the Christian faith, will be absent in theirs. The sun, moon, stars which obviously represented a physical place but which clearly could not be journeyed to by any human means (at the time when Christianity was being formulated), therefore arose as the site to which we journey after death.

In most religions, daily behavior is regulated by the hope of some form of reward, either as rebirth, reincarnation at a higher level of consciousness, joining with God, or some other ultimate fulfillment. In the case of Christianity, desired behavior is motivated by the perpetuation of belief in a positive after-death residence in the heavens or punishment by the banishment to the underground world - for indeed there has always an assumption that below the ground, existed something. That something was even more unknown (and in many respects still is) than celestial space, for at least people could see into the open reaches. Probably for that reason, the most negative associations were assigned to the most unknown areas, the underground. This is reasonable, given that so many of the images and experiences associated with atmospheric phenomena are pleasant, life-giving and positive.

The one aspect of the mythology of Christianity that turns out to be completely correct, is that the deep interior of the Earth is hot. This fact was probably deduced from the observations easily made even without advanced technology; volcanic sites spewed or oozed molten rock and sulfurous stench and in other places steam and hot water sometimes sprayed from the Earth. These clues probably contributed to the location of hell being described as deep under the Earth.

A rather different universe might be pictured in the beliefs of the underground religious. They would most likely hold some god or creator as the centerpiece of their system, for most religions (and the Judeo-Christian tradition is no different) feature the acceptance of an omnipotent figure as the absolute pinnacle of sentient existence, and an earthly, human intermediary; a scholar, philosopher, messenger, prophet or the like, to interpret, codify and bring the will of the god to the people. Their religion may follow the same pattern. However, with a closed universe their only experience, heaven, or at least the source of the creator may be the water or ocean. If bodies of water extend into occupied spaces and especially if it is the source of food and breathable air, it is reasonable to expect that their worship would center around some features of the life-giving sea. It would represent life, freedom (being more easily penetrated than rock), and the source of all that sustains them. They well may imagine that heaven is an area of unlimited water which they cannot reach in their lifetimes but in which they reside after death. Water, for them, may represent the same medium as space does for us.

By extension, their religion may offer the hope of reincarnation into some form of sea life as a higher level of existence. Their god may be a sea god. There is precedence for this in the ancient Roman and Greek myths of Neptune and Poseidon. Burial rituals in such a religion may include consignment of the body to the sea. Land burial is reasonable for us since life-giving crops sprout form the Earth our livelihood depends on the soil.

Another possibility, repugnant to us, is that cannibalism is a part of the ceremonies which mark the end of life. If the food chain in their environment is a limited one and sufficient nourishment is a social problem, then a socially-supported ritual such as consuming part of the dead would be useful, at least symbolically, in encouraging ecologically sound practices. It may be that this is further reinforced by being associated with respect and ultimate fulfillment by joining with another living body, and in that way, continue to live. This idea may be symbolic or thought to be actual. Many may expect to experience renewed life in consciousness in the joining with another, while others see the practice as only an aspect of the faith which comforts and promotes a cultural bond.

Even among the faithful followers of our religions there are strongly held differences of opinion regarding the literalness of scriptural dogma.

If their natural environment features any geothermal sites that have open magma flow, it may be that these are, or have been, worshipped or given special holy significance in their religion(s). This would be entirely reasonable, since molten rock would be the most intense form of heat known to them and would represent the concentrated energy which makes their existence possible. Such sites may even be considered shrines and be the center of religious proceedings. This would bring about their functioning in a curious dual role which would seem to us unusual: These energy-radiating openings to the inner Earth might function as holy places as well as centers of manufacturing. Religious and industrial leaders might share an interest in preserving and exploiting the symbolic as well as the practical potential of these geological phenomena, for religion is not only a system of personal spiritual beliefs, but also a means of social control and an arm of the ruling structure.

Alliances between the church and the state have been at various times and in various cultures, features of our social organization. In fact, it is the competition for control of the loyalties and behavior of the population (not to mention the revenues) that has been a source of conflict between governmental and spiritual rulers. Resolution of this natural conflict has led to different solutions in contemporary societies: Most western democracies and democratic socialists nations have stripped the church of many of its ruling powers by establishing, to some degree or other, a separation of powers between church and state. Communist nations have diminished the church's powers even more by usurping many of the functions of religious leaders and abolishing certain others.

There is undeniably a reciprocal relationship which exists among the major institutions of power within a society, be those autocratic rulers, elected officials, business organizations or religious orders. The balance of influence among these groups shifts over time to conform to changing conditions and attitudes. The changes are both a result of attitudes and shape attitudes as well. Consider that few people in our modern world, regardless of their faith, would willingly accept the religious control over their lives that was exerted by church officials in past centuries. As one institution gains more control of individual loyalties, and increasingly assumes the role as the primary governing body in society, the power of the other institutions recede. This shifting in power towards and away from various institution has been going on since our earliest human organization and is still continuing.

After the rise of Christianity in the western world, the faith was modified to conform to modern thinking and given regional identities to allow it to function as an integral part of the cultures to which it spread. But as a central doctrine which defined the parameters of allowable human behavior, it dominated religious thought in the western world and became the most powerful force for social control the world had ever known. This was not the accidental byproduct a good idea which caught on, for the active and sometimes even forceful conversion to Christianity of peoples without Christian beliefs has been a feature of all branches of Christianity practically since its inception. It is little wonder that rulers - kings and commissars alike, had reason to subordinate the church's power.

In most industrialized western nations, a relatively new and powerful force has risen to take the place traditionally occupied by the church in maintaining social control; that is the business/financial community. The incredible rise to power of the large multinational corporate entities represents the most phenomenal growth of any social order in history. Although their dominion is not the spiritual life of individuals, but the financial and temporal aspects, they represent, taken as a whole, potential for control over populations that rivals or exceeds that ever held by any religion. And it is clear from the conservative political shifts that have taken place in the governments of nearly all industrialized countries, that these institutions are influential indeed. The power of persuasion and manipulation over attitudes and loyalties employed by these modern institutions, given their dominance over entertainment and communications media, is far more sophisticated than the techniques traditionally used by religions.

Just as has been the case with religious organizations, there is little that is democratic about business organizations. Since they have grown and developed in an organic-like process that is somewhat analogous to natural selection (if fact business leaders themselves often use the analogy of "survival of the fittest" to describe the business environment), they feature a dominant-subordinate organizational structure that is reminiscent of tribal, and later autocratic regimes.

While this alliance between financial business interests and governments may be new to us, at least to the degree that it is now practiced, it may have been a feature of a subterranean culture since the beginning. It may be that a close cooperative relationship exists between church and state or between church, state and industry to rule through religious beliefs, governmental order, and financial organization. Theirs may be a social order bound by the mutual ruling of religious, business and governmental institutions or by any combination of these, for if adequate social control can be maintained by religion and business interests alone, there would be little need for other forms of government. In their society, elected government may play the subordinate role that religion now plays in ours.

EXPLORERS

The urge to explore is as old as recorded history and presumably existed long before that. An underground people would have almost surely followed every passageway and examined all interior chambers which they could access. What then? Would they have assumed that there was no more to the universe than they could find? Probably not. Man is by nature curious. Myths and legendary accounts of creation, with appropriate gods to explain, not only the physical world, but in some accounts, the behavior of people and animals as well, have always been a part of our cultures. Since early underground humanoids would have been our distant ancestors, they likely would have fashioned a belief system of some sort to explain the conditions in which they found themselves - the world as they knew it. We must assume that their thoughts included some curiosity about what lay beyond the walls of their home. And curiosity is not the only motivating force for exploration. The prospect for personal or collective gain, the need for more space or the acquisition of new possessions and to increase security have all been additional cause to venture out to unknown places.

After exhausting all accessible locations, these creatures must have wondered if there existed, somewhere else, more places like the ones they knew. Further exploration for them would have been a matter of digging, drilling, tunneling and perhaps exploring underwater in an attempt to expand their knowledge and their resources. With a premium on space, available materials and food supplies, they may have been forced into a desperate search for more living room. It may even be that disturbances caused by our geologic explorations, or our underground nuclear testing provoked their concern and prompted their search for the source of the shocks.

As their technological means improved, their pursuit would have become more productive and perhaps, in time they broke through not only to additional underground openings, but perhaps finally to the surface. If that is the case, it must have been a shock that defies our understanding. An entirely new dimension of existence would have been revealed to them, the implications of which, their philosophies and explanations would not have prepared them for. It would be like our discovering walls at the edge of the universe!

For them, outward exploration would have held greater peril than space travel has for us. Our space flights are risky for the astronauts but do not pose much danger to our home planet (although there was serious concern about the possibility of infectious microorganisms being brought back to Earth which caused the cautionary quarantining of the earliest space travelers). If subterranean explorations were in the form of penetrations through solid material, they may have hit magma, oil, natural gas or water, any of which could have caused the disastrous destruction of much of their world, or at least caused considerable damage. Perhaps they had such mishaps and suffered serious consequences as a result of their daring. Perhaps this inhibited further exploration. Perhaps they imposed moratoriums on further expansion or changed their methods. Possibly they chose instead to dive as far as possible beneath bodies of water in search of the outer world. Perhaps finally they explored successfully, ultimately breaking through to the great open spaces of our world!

The different patterns of audio reflection must have completely confused their hearing or echolocation sensors. Their vision, whatever form it takes, could not make sense of the emptiness above. The startling "differentness" they encountered would have inspired much caution, for this would be unlike anything ever experienced. The prudent approach would be for them to cautiously expand their search. Expeditionary forces would have made tentative incursions into our environment to attempt to gain as much information as possible with minimal risk. These forays would logically take place at night, circumstances most comfortable for them, for sunlight would be extraordinarily dangerous to such creatures. They might be almost as sensitive to bright sunlight as the legendary vampires, shunning daylight completely.

Depending on where they emerged, sooner or later they would have detected animals, signs of civilization and human beings. The implications of discovering creatures so like themselves would have taken years or decades of discussion to reconcile and plan appropriate action. It's likely that these creatures would have considerable fear of us. To them, our environment would seem terribly frightening; the harshness of daylight, the empty black overhead of night, the sheer physical size of everything - including us, the tremendous energy which is evident everywhere in the movement of massive objects, the incredible expanse of land and seas - and the sea of people. Our civilization would seem huge and energetic beyond imagination and soon they would learn that we possess weapons of monstrous destruction, with the accompanying will to use them. They would have studied us, surreptitiously, for as long as possible before initiating contact, if at all.

How would they get about on our terrain? Would they be limited strictly to foot travel or would they have a means of transportation? If their technology has advanced far enough to have some kind of artificial locomotion, that is surely how they would explore our world. Distances are so great, with oceans of water to cross, some vehicular craft would be almost a necessity, especially if the assumption is correct that daylight conditions are too harsh for them. Extensive foot travel is physically taxing and would require shelter from solar exposure during the days. The risk of being discovered is so great as to be a certainty if traveling in highly populated areas. A craft would allow for quick retreat if seen, and provide security as well as a place to store necessary supplies.

When we undertook to explore space, it was necessary to build special vehicles particularly suited to the task. None of the methods of transportation which we normally used was up to the unique requirements of hurtling through the atmosphere at extremely high speeds, as happens upon reentry, and withstand the tremendous temperatures that result. New materials had to be created, for weight was also a paramount consideration.

Efficiency, fuel consumption, physical size, heat resistance and many, many other factors distinguished the spacecraft from earthbound vehicles. When our exploration extended to the surface of the moon, a special vehicle again had to be designed, and although it somewhat resembled a common car or truck, it still had to be transported through space and function in an environment with no atmosphere and much less gravity than on Earth. Much of its familiar appearance was superficial, for it too was uniquely suited to lunar exploration.

Whatever methods of powered transportation exist for underground people, they would not likely serve for surface exploration. In caves, surfaces are too irregular for conventional wheeled vehicles and open spaces are too limited for extended flight (except by devices of neutral gravity and extreme maneuverability). However, navigation of the air and seas would be absolute requirements for exploring the surface. We have not had any real need to create single vehicles which can function in air, water and on land, but such craft can certainly be built. We have found it more practical and economical to simply use different systems of transportation specifically designed for the medium. But if one had to traverse all the different mediums found on the surface of the Earth, and had not the luxury of several types of transport, a single all-purpose craft would be the logical solution.

In order to evade detection (an especially urgent concern if exploring a potentially hostile environment), travel in the air and sea would be more important than land travel, as moving about on the land requires prepared pathways, slower speeds and surely would be noticed. Sea travel would also be required if emergence from the underground had to be made through water. Such a craft would likely be quite small by our standards, holding only a few aliens. It would also need to be highly maneuverable. If weapons could not be used (and they could not, if clandestine operation were the goal), speed, maneuverability and small size would be the best defense. Reports of sighting can be disbelieved, damage to life and property cannot be denied and would also provoke a search and possible reprisals.

Another possibility exists, and that is that just as we use multi-stage rockets to lift spacecraft into orbit which can then land like more conventional aircraft, or just as we use aircraft carriers to transport planes to offshore locations, subterranean creatures may use one craft to leave their underground environment from under water, carrying a smaller, lighter airborne vehicle to explore the skies and lands, which returns to the aquatic "mother ship".

We cannot assume that their exploration of our world would be by vehicles that resemble their normal modes of transportation any more than our explorations have. After all, we would not try to go to the moon in a pickup truck. Surely, upon discovering the nature of the surface, special craft were designed solely for the purpose; craft which allowed them to make quick, secret visits to various places and return to their world undetected. Although they would be exploring the surface world, it would involve mostly air travel and ironically, if they were observed the assumption would most likely be that they originated from above, not below.

Communication with us would not only be risky and unwise, initially it would be impossible. Deciphering our languages would be extremely difficult for them, even more difficult that our translating ancient dialects, for not only would the symbols be different but the manner of recording itself would be entirely incompatible, as already discussed. Our written or spoken language would first have to be converted into a format which they could sense before study could begin. And even after an intellectual understanding worked out, it is doubtful they could make the sounds needed to reproduce our speech patterns. The differences in written formats between their language and ours would be barriers to nonverbal communication, so formal contact would be postponed for as long as practical. In the meantime, observation would be the best means of gaining enough information to prepare them for encounter.

The temptation to ignore the surface and retreat to the relative safety of their own environment must have been great. In numbers and technology our world would be overwhelming. Contact with us would be very dangerous indeed. But it would be bound to happen anyway, if not initiated by them, we would discover them sooner or later. With continued drilling, mining, more sophisticated inner-earth sensing methods, increasing undersea investigations, robotic tunneling, mohole-type projects, and so forth, one day we would find them - and if they were unprepared, with no understanding of us at all, they would be at a distinct disadvantage. If they could at least learn some of our language, they would be much better off than us when contact is first established. The clock then, was running, and ignoring us would not be wise at all.

Their need for detailed intelligence about us would ultimately require more information about the physical nature of human beings however, and would have finally required contact. Of course contact would have to be limited, and since no linguistic communication would likely be possible, abduction would be the only course of action.

THE EVIDENCE

The most compelling evidence for the present-day existence of subterranean humanoids is the persistence of reports of contact by humans with creatures of unknown origins, and that if true, are more readily explained by this hypothesis than by an extraterrestrial one.

The phenomenon of contact with non-human intelligences is not a recent one. For centuries people have reported visions, voices, communications or other kinds of personal contact with creatures of an unearthly origin. These have been characterized by descriptions consistent with whatever belief system or mythology that was the basic mystical paradigm of the individuals reporting the event. In other words, the alien description was framed in a cultural context. That alone, gives some suspicion to the twentieth-century reports of spacecraft and aliens of extraterrestrial origin. Clearly, the advent of aircraft, rocketry, increased knowledge of astronomical bodies, and the prospect of space exploration by our own species has caused the belief in non-human intelligences to be couched in terms of space voyagers from other planets. Many, if not most such reports are obvious hoaxes. They display the unsophisticated assumptions about interplanetary and intergalactic travel that are clearly the product of science-fiction movies and books.

Still, there are some troubling contradictions in considering such reports. We are faced with the situation in which we must view every single one of the reported incidents as a product of human frailty and imagination or believe that there are a few, a very few, which form a core of truth and cannot easily be dismissed. The danger in contemplating any issue in which the predominance of information is useless, is that the entire issue can be too easily be dismissed; that the mass of misinformation will induce a bias that obscures the possibility of authentic experiences which deserve serious consideration. It is difficult not to allow the cynicism of experience to poison objectivity.

We are accustomed to accepting testimony of witnesses as evidence for all kinds of purposes. Despite the obvious errors, exaggerations, lies, misinterpretations, hallucinations, dreams, etc.,etc. to which human beings seem prone, we are, on the whole, fairly reliable, capable and observant. In all areas of our daily activities we trust our well-being, safety - our very lives and those of our loved ones to the perception and judgment of our fellows, and it serves us well, for the most part.

What are we to do then about reports of contact with unusual creatures? Are we to dismiss the entire matter? Can we say that every such case is without substance? If we allow the possibility that even a very few cases of reported contacts contain some truth, even if inaccurately reported, then should we not look for a viable explanation? If so, the usual explanation that such aliens are from outer space is the *least* plausible one. And it is the lack of a plausible alternative that causes us the greatest skepticism. But just as it is foolish to assume that simply because we may not have a known explanation for an unidentified flying object or be able to imagine one, it must be a space alien, it is likewise foolish to assume that because the space alien explanation is highly improbable, the report must necessarily be false.

In August of 1955, in Kelly-Hopkinsonville, Kentucky, eleven members of a single family reported seeing several unusual creatures outside their rural home. The terrified family included eight adults and three children, all of whom observed the creatures intermittently over a period of many hours, from 7;00 P.M. to 5:00 A.M., while hiding behind locked doors. Several shots were fired at the creatures but none apparently hit their mark. The incident was reported to and investigated by local law enforcement authorities who found no ordinary explanation to account for the witnesses' observations.

In cases such as this, fraud is always a possibility, however in this instance, as in many others, it seems unlikely due to lack of motive. The individuals in question had nothing to gain from fabricating such an incident, and in fact suffered considerable ridicule as consequence. Of course it is possible they were the victim of a prank, however given that the creatures were fired upon with rifle and shotgun, yet still returned, makes this questionable. Misinterpretation is the other obvious explanation, however given the length of the observation and the number of witnesses, this too seems improbable; the description offered by the witnesses is so unique that confusion with any type of animal native to the area or humans seems hardly possible.

The appearance described by all members of the family was of humanoid creatures approximately three to four feet tall, with very large heads, huge glowing eyes and huge flared ears. The creatures were also reported to be hairless and to glow in the dark. The arms were described as extremely long with very large, claw-like hands.

This physical appearance is completely consistent with the probable features of a subterranean humanoid as described in chapter 6.

The only feature which is not a part of the expected profile is the reported glowing in the dark. However, many forms of marine life which inhabit the deep ocean trenches, where there is absolutely no light, do glow in the dark. In fact, three-quarters of swimming fish found at depths below 10,000 feet exhibit some form of bioluminescence. This allows recognition of other members of like species (just as do distinctive markings) for mating and also serves to confuse or disorient predators. So if we think the reports of humanoids that glow are simply too bizarre to possibly be true, we must remember that there exists biological precedent for this as well as all of the other features described.

There are many documented reports of encounters with alien creatures and it is not the purpose here to recount these. Published accounts are widely available, though of varying repute. I cannot and will not in any way validate the one cited here, nor do I offer it as one of the most likely authentic. It is included as typical of many, and to illustrate the contradiction between the seemingly honest testimony of multiple witnesses and the seemingly illogical or impossible circumstances described. It is this conflict in the absence of a plausible theory that inhibits serious investigation. The subterranean hypothesis would reconcile many of these contradictions.

Close Encounters

Close Encounters of the Third Kind is the term for reports of direct human contact with aliens. There is no way of proving or disproving the accounts. Many have been investigated at great length, though it must be said that, for the most part, the investigators have been individuals who have made a career of investigating and writing about such incidents, not objective scientists. Books published on the subject which tend to legitimize the stories are seldom objective and unbiased. Lurid television programs, masquerading as journalistic or investigative, do nothing more that exploit the mysterious nature of the subject to attract viewers and advertising dollars (people who are easily manipulated by television are the advertising industries' delight, and so shows which can command a large audience of "believers" can command premium fees for advertising air time). Information contained in such books and broadcast programs is generally useless.

It may be that literally every reported case of contact or abduction is entirely false; each being the product of different and unique circumstances that can be accounted for by other explanations. And in this connection it should be pointed out that simply because no one has an acceptable explanation for any given situation does not mean that one does not exist. The typical comment of uncritical thinkers who cite, as evidence to support the theory of UFOs in conjunction with extraterrestrial aliens, "what else could it be" miss the point. The failure of imagination is not proof of anything. Professional illusionists routinely create effects for which few in the audience can provide any explanation. Yet on the rare occasions when the secret method is revealed, we all respond similarly; "of course, I should have known". And this failure to discover the trick on our own does not lead any of us to believe in magic.

The greatest barrier to our acceptance of the testimony of witnesses in extraordinary cases is that the reported incidents are so unbelievable that surely it is the witnesses who are mistaken, hallucinating, dreaming, mentally unstable or perpetrating a hoax. If the statements made by most of these individuals were of something less fantastic, something we know can occur, most would have been believed without a doubt. This is, of course, a normal and even appropriate response to extraordinary statements without physical proof. We relate all new information to our own experiences.

However we must also temper our estimation of events described by others not only by our own experience, but also by our sense of reason and scientific knowledge. We doubt the reports of alien beings because we think it cannot be so. But as we have seen, and continue to see in science, ideas about what can and cannot be so keep expanding. We know that unusual adaptations of life can exist here on our own planet, never mind the question of life in other galaxies, and we have reasoned that it is also possible that there can be unknown forms of humanoids from unknown parts of our own world. We must reevaluate the human accounts of such phenomena in light of the possibility that such encounters are not with beings from distant worlds but with mutated members of our own species.

Of the accounts available from the few serious investigations of reports by witnesses, a few consistent features are interesting: In almost every case, the aliens appear humanoid but significantly smaller than we, typically, three to four feet in height and slimmer as well. The creatures' heads are almost universally described as large and having no facial hair. Skin is usually characterized as pale or grayish and the eyes, large and dark, sometimes slanting. This appearance has come to be a rather stereotypical image of aliens as described in what seem to be the most reliable accounts of abductions and close encounters of the third kind. I make no judgment about the authenticity of such reports, but simply point out that the descriptions are not inconsistent in any way with the appearance we would expect in subterranean humanoids. Exploration of the surface world by beings from within seems at least as plausible as the hypothesis that we are being visited by creatures from extraterrestrial sources who, just coincidentally, resemble us in so many ways.

There is another consistent feature of reports that is puzzling: In most cases in which abductees recount communication with aliens, the aliens express concern over mankind's use of the planet, describing ecological catastrophes which may befall us if we do not alter our ways. In some cases the aliens communicate that the actual purpose of contact with humans is to warn us of the peril we face by following our present practices. That an outside intelligence might warn of ecological disasters if we do not prudently consider the biological and ecological consequences of activities which impact ecosystems is not hard to understand. Experts in various Earth sciences: geologists, meteorologists, oceanographers as well as biologists, anthropologists, sociologists, psychologists, political scientists, journalists and virtually all environmentalists have been saying the same thing for decades.

What is puzzling is; why would extraterrestrial beings care? Even if the Earth were to explode in a violent apocalypse, it would be of no consequence to them. If fact they probably could not even detect it from their planet. Things explode in space all the time and on a far more massive scale than any disaster that might occur on our tiny world. Exploding things in space are as common as crabgrass; explosions are to intergalactic space what lightening is to the atmosphere. Are we to assume that some alien culture went to all the trouble to travel thousands or millions of light years simply to warn us that we may be irrevocably damaging our planet? Could they be that benevolent? Would one planet and its lifeforms be worth that much to them? How would they have known this in advance of their visit? Did they set out to reach us specifically or did they just happen upon us in a process of general exploration? And if they were from a far-away place and went to the considerable effort to reach us to tell us about our errant habits (even if they could have known, which they could not) why not just come right out and openly contact us officially, make a formal visit and announce their presence and purpose to all so there could be no doubt about their existence and their purpose? Surely this would be more effective in motivating us to alter our practices than vague suggestions reported by individuals whose unusual experiences serve to discredit them.

One answer to this might be that the intentions of the visitors are not benevolent. They may be intent on using the planet and us for their own purposes at some later time; that they may really be bent on conquest. However this still does not answer the question, for if the expressions of concern over our environment are merely propaganda to cover their real mission and at the same time influence our behaviors to serve their agenda, an all-out public relations campaign, complete with visiting emissaries bearing information and technology which we could use to preserve the planet would more effectively modify governmental and industrial policies than would clandestine contacts with lone individuals who have no political influence. Surely revealing themselves to us would carry no risk. We do not have the capability of reaching them, nor will we in the foreseeable future, and to protect their security they would simply need only to tell us they came from somewhere far from their actual home. We could not possible ever detect the deceit. We could not track them through space and learn to where they really return. Their home planet would be completely safe from any intrusions by us. So why not just come right out and openly contact us?

Of course the obvious answer is that such communications are entirely the product of the imaginations of the abductees. The communications are usually reported as telepathic in some way. The abductee typically describes the experience as "somehow knowing" what the alien is trying to communicate, that the voices reveal themselves in the mind's ear. This form of communication, while certainly not beyond the realm of possibility for creatures who may posses faculties which seem to us fantastic (in fact is very likely that they would seem fantastic to us - it is the very similarity to us that makes such human-looking and behaving creatures so hard to accept as extraterrestrials) seems suspect. The reported telepathic nature of the communication rather obviously suggests that it does not take place at all. Even if the reports of the contact and abduction are true, the communications could still be simply the projection of the victim's own thoughts onto the alien.

A third possibility exists. That they are not from an extraterrestrial source but represent a subterranean species, carefully making initial, cautious contacts with those who control the majority of the planet. In this case, the reluctance to reveal themselves to us is completely understandable. They would be at considerable risk if we knew of their existence. Without knowing our intentions, they would be very reluctant to make their habitat known, yet this could certainly explain their concern for the well-being of this planet. Their vested interests in the health of planetary ecosystems may be even greater than our own. Pollution of the oceans, radioactive contamination, underground burial of nuclear waste, possible contamination of deep underground water, which may effect us may be catastrophic to them, for surely their ecosystems are smaller and more delicately-balanced than ours. They may have the most urgent need to get the message to us without arousing any suspicion about their presence. In this case it would be entirely in their interests to allow us to believe that they originated on some distant planet.

It is natural for us that our exploration for other forms of life be interstellar in nature, for the heavens have been our preoccupation. Since we have lived and evolved under the mantle of the stars, we have speculated about their nature since earliest times. The heavens have dominated our lore, made possible ocean navigation, figured prominently in our religions, provided invaluable scientific information about the nature of the universe and matter itself, and has become the assumed home of any undiscovered forms of life. We may be surprised. Our first contact with aliens may be with "superterrestrials" and not extraterrestrials. The history of our explorations has been outward to the surrounding surfaces of our planet and then, as technology has permitted, upward to those areas that have so long mystified us. Perhaps it is time for us to look in other directions.

EXTRATERRESTRIAL SPECULATIONS

SETI (Search for Extraterrestrial Intelligence), a project to listen for evidence of other intelligences in the universe, demonstrates how seriously space scientists take the possibility of intelligent life on other worlds. Given the demonstrated ability of life on our own planet to adapt to environmental extremes, perhaps equally serious consideration should be given to the prospect of advanced life from within unknown areas of Earth.

So much of the study of the underground itself has been motivated by economic considerations directly and indirectly: The exploration, mining and drilling for various ores and substances which are indispensable for virtually all forms of economic development, the ever growing demand for fresh water, even the millennia-old work of agriculture require study of the Earth, but have dictated and narrowly-defined the nature of the exploration.

Only recently, has purely scientific study begun, and still not with the fervor and commensurate expenditures as the exploration of space. Paradoxically, expanding economic development with the increasingly urgent need to find reliable methods of earthquake prediction are forcing much of the current effort to better understand the Earth.

The topic of aliens, UFOs, anecdotal reports of abductions by unknown creatures, etc., is so fraught with poor information, little or badly done research, stereotypes, stigmatization and almost no validated physical evidence, that is hardly any wonder that it is largely avoided by serious scientists. Who can blame anyone, with a living to make, for not jeopardizing their careers, professional reputations and their ability to get a job by involving themselves in such a controversial issue. It is probably largely for this reason that most of the books and materials on the subject are the work of writers seeking to cater to a market for sensationalist literature or of amateurs who are seldom unbiased. I hesitate to even broach the subject, however the context of this speculation demands that the thesis be articulated to its logical conclusion.

If, and with the strongest possible emphasis on IF, there is any actual case of visitation by an alien intelligence, it may be as likely, if not more likely, that their origin is from within our own planet than from a stellar source.

I must emphasize that this is not a discussion of the existence or nonexistence of alien visits or presence in our mists. Such an examination would require volumes, and I fear that sources of information and evidence have now become so contaminated that serious investigation is likely impossible. Insofar as whether we have been or will be visited by unknown intelligent life forms will have to wait for incontrovertible evidence, or continued lack of such.

But for the sake of rational discussion let's consider some biological realities of terrestrial life as opposed to possible extraterrestrial life. Earthly creatures, considering just land animals alone, or even land *mammals* only, display such diversity of size, color, shape, appearance, vocalizations, habits, and so forth that almost none except our primate relatives resemble human form at all. Yet, if there is one trait that is commonly shared by practically all reports of contact with aliens, it is that they are "humanoid" in appearance. And if we consider the incredible diversity of all forms of life on our planet, including all those of the deep seas and the air as well, none resemble us at all. And these are all creatures with which we share similar biochemical functions as well as a home planet.

It seems more than far-fetched that creatures from otherworldly places, apart as far in time as in distance, and from environments that are surely quite unlike our own, would be at all humanoid. Given the alleged appearance of aliens, it would be more likely that they are from some region of our own planet and are, in fact, of the same ancestral origin as ourselves. Subterranean or suboceanic locations are about the only earthly sites possible. The subterranean hypothesis would reconcile many of the conflicts which make the possibility of alien contacts rather difficult to believe.

Scientists who are skeptical about earthly visitation by extraterrestrials point out the almost incomprehensible distances in space and the span of time between ourselves and even the nearest potential sources of visitors. The frequency of the reports is also a confusing factor; If aliens came so far, would the visits be so frequent and would not the length of their stay be longer?

Also, the rather earthly appearance of the reported crafts is troublesome. Explorers of the universe would not likely be so prepared to navigate in an atmosphere exactly like ours given the infinite variety of conditions they might encounter on their travels. The craft often described seem unlikely to have originated on other galaxies or be capable of such extensive voyages. Interstellar space is filled with dangers: radiation, meteors, incredible heat, dust and other particulate matter which could wreck havoc with fragile craft - and the longer the voyage, the greater the risk and possible damage.

In addition to their similar appearance to ourselves (two arms, two legs, torso with one head, two eyes, erect posture, and so on) they are reportedly near to our own stature; an extremely unlikely circumstance. Even if there were something about the development of intelligence which made human form necessary or probable (which it doesn't, just look at porpoises as an example), why might they not be fifty feet tall as well as five feet tall, especially since their gravity would surely be greater or lesser than ours. For these and other sound reasons, the reports of extraterrestrial aliens on Earth is almost universally rejected out of hand by scientists.

It is entirely possible that the reports are all the product of distortions of imagination, and that the descriptions are such because of our human trait of creating our gods (and our visitors) in our own image. The stereotypical interplanetary alien has become well-imbedded in popular culture. Police investigators are tiresomely familiar with "copycat" crimes that are committed after widespread publicity that surrounds sensational initial crimes saturates the media. Multiple confessions are also a part of the pattern of public behavior that accompanies notorious events.

Perhaps, alien and UFO sightings are simply another manifestation of this human phenomenon.

It is worth noting though, that in criminal courts, defendants are routinely convicted and sentenced to long prison terms and even death, solely on the testimony of witnesses and with less hard evidence than is available for the existence of alien life on Earth.

DISCOVERY

Some interesting parallels apply between our first steps at exploring space, especially our search for intelligent life, and an underground peoples' exploration of the outer world. As already touched upon, we have undertaken by a variety of means, attempts to determine whether intelligent life exists on other worlds and to contact, or at least make ourselves known to them. Methods of detecting intelligence outside Earth has taken the form of listening and watching for signals in the form of visible light, radio signals, or other frequencies of the electromagnetic spectrum, that form a pattern which suggests that others are attempting to make themselves known, or which at least do not seem to be random enough or regular enough to be of natural origin. This approach is essentially not too different from traditional astronomical observations. It is relatively inexpensive and there has been little controversy associated with this kind of activity.

However broadcasting signals or sending physical artifacts adrift to make ourselves known is another matter. Some feel that announcing our presence, and especially revealing our location to life forms about which we know absolutely nothing, is a risky move. Suppose the information is intercepted by aliens of far superior intelligence and technological means. Would they not view us as insignificant and expendable as we commonly view insects? We intuitively feel that our "humanity" should hold us in good stead with any other intelligent life form, especially one which has the capability of traveling in space. Likewise, it is rather optimistically assumed that a civilization that has conquered the technical and social barriers of escaping their native planet would, of necessity, have advanced beyond destructive behaviors. Are these reasonable assumptions?

Of course we have been inadvertently sending signals into space since the introduction of electric lights and radio and television communication. The illumination generated by electric light sources is rather puny and probably cannot be seen very far. Radio waves are spreading out into space and may be detected farther away, but are likely indecipherable and so weak as to provide little substantial evidence of our existence. So at this point we probably have done nothing too dangerous. But as our technical means advance, and the use of lasers and other increasingly coherent and powerful signals are employed, it becomes an important concern.

How would we react, or rather how will we react when indisputable evidence of intelligence is located on some other world; for many in the scientific community feel that the discovery of intelligent life in space is only a matter of time. Are we at risk? Should we immediately make ourselves known or lie low for a period of time, researching the possibilities and attempting to learn all we can about them and their potential. It may be a moot point. The distances and time which separate us will almost surely make any discovery a matter for our most distant descendants to deal with, if at all. Unless some remarkable features of the universe are discovered which are now only science fiction, it's likely the evidence we encounter will be more archaeological in nature - remnants of a long past civilization, as isolated by time as by distance - than signs of impending intergalactic intercourse. And by the time any other intelligence detects us, we too may be long gone, or at what's left of us moved elsewhere.

The same concerns must have faced underground people. Did they explore first to better their understanding of the environment or expand their domain? Did they wonder if other life, especially intelligent life, existed elsewhere? Did they have the same anxieties as we? Do they also assume benevolent traits in unknown life forms?

For them, the search for other life may have been the result of investigations into some unknown phenomena which they detected, caused either by our explorations or military activities. As already discussed, their approach would almost surely be as clandestine as possible. But how should we search for them, if we decide to? Do we just wait for them to make themselves known to us, or until some incontrovertible physical evidence appears? Should we not at least consider the possibility of their existence as seriously as that of interplanetary or intergalactic life, and make some attempt to learn what we can before they reveal themselves? Do have anything to fear?

We probably do not have much to fear, at least in the traditional adversarial sense. We surely outnumber them and our war machines are quite effective. There exists the possibility that some other means of destruction, one for which we are poorly prepared, like biological or chemical weapons, genetic sabotage, terrorism, bacterial or viral epidemic or the like could be employed against us if they chose. But this seems a remote scenario. They would not likely find our environment easily habitable or attractive.

Perhaps we should make some attempts at detection purely in the interest of science and curiosity, or even good business. For surely this nation's entrepreneurs can find ways to expand economic development out of a discovery of such magnitude. In any case, how would we best go about looking for them?

Methods for detecting underground intelligent life fall into three categories: direct exploration, remote sensing and indirect monitoring. The first approach is self-explanatory. It would involve drilling in areas considered to be likely sites, with subsequent investigations through the other two means if promising evidence resulted. Direct exploration is rather costly and time consuming, and to some extent is a hit or miss affair. Very deep drilling is quite impractical in many of the areas which might best warrant study. Remote sensing might involve sending video cameras, audio devices and other information gathering sensors down into openings breached by boreholes and observing via radio transmissions to the surface. This would be the only practical way of directly looking into any caverns discovered.

The best approach, at least initially, would be to listen for distinctive sound transmissions which may be the result of intelligent life. Solids and liquids are better transmitters of sound waves than air, so whatever mechanical shocks are induced by any underground activity are detectable at some distance. The problem is that the underground is a very noisy place, at least as far as low level sounds are concerned. There is lots of movement going on, some of it quite massive, which generates enough noise to mask the goings-on of any living creatures. Added to this is the fact that artificially-created sounds would meld together and be difficult to discern as anything out of the ordinary even if they could be distinguished form the background.

This is where computer technology is useful. Sound is transmitted by mechanical waves through a physical medium, unlike electromagnetic radiation (visible light, radio waves, X-rays, microwaves, for example). Sounds which are below the threshold of hearing, or even instrumental detection, can be enhanced by the addition of more sound in the form of white noise (a mixture of all audible frequencies) or some band of white noise. Sounds of the same frequency which are in phase, that is those whose waves are completely synchronized, will reinforce one another to strengthen that particular frequency. When two such waves are completely out of sync., that is 180 degrees out of phase, they cancel each. Therefore a blend of frequencies can be overlaid a signal which is too weak to be detected, and the reinforcement effect of complimentary frequency and phase will strengthen the latent sound to detectable levels.

But detection of mere sound is not, by itself, enough to indicate an intelligent presence. Patterns must also be discerned. As with the effort to detect signs of life in space, underground monitoring also has to search for orderings of sounds that are neither too regular nor too random to be of natural origin. In this particular case, it is extremely unlikely that aliens are beaming out any message with the hopes that it will be picked up by a friendly planet (given the range of such broadcasts, the risk would be too great), so we cannot expect to hear binary numbers which represent mathematical equations or atomic structures, or a series of prime integers which would be indisputable evidence of an intelligence. Here we must try to find the unintended noise that is the by-product of technology; sounds generated by transportation, building, power generation, environment shaping and so forth.

Computers can be programmed to methodically scan for specific frequencies and to search for certain patterns in the sequence of those sounds, keeping track only of those desired findings which can later be reviewed and interpreted by humans. In this way, unmanned listening posts could be established in locations unsuited for permanent staffing. This would be a relatively cost-effective way of monitoring many areas.

This will not be easy to do however. The sought-after signals will have to be sorted out from the sounds of moving water, grinding rocks, flowing magma, fault shifting, not to mention the more than 700,000 earthquakes which occur worldwide each year. Even at best, such a project will be little more than sophisticated guesswork and could not clearly establish alien presence underground. It might however provide clues, along with anthropological evidence and theories, as to the best places to start the search.

EPICENTER

Exploration deeper within the Earth is not only less glamorous, but in many ways more difficult than space exploration. To date, the deepest marine drilling has extended to a depth of only 5,480 feet below the sea floor with the deepest penetration from land surface being 31,441 feet. The deepest mine is 12,600 feet while the deepest known caves are only a mile or less below the surface. Probing underground is a more blind affair than probing outer space. The extreme pressures and the hardness of many layers of rock permit only small orifices, and information obtained from drilling such small peepholes, especially at deeper depths, may not yield much information about conditions even a few feet from the actual hole.

If we can imagine attempting to explore the surface of the Earth from within by drilling tiny holes to the outside, we can see that very little actual information about our life and civilization would be obtained. It is probable that even with many thousands of holes, no evidence of beings would be found, especially if we were not looking for it. And this, in spite of our having a world population that we consider to be enormous. The exploratory penetrations would most likely break through to the surface under water, under ice or in an entirely uninhabited area, or even if the probe happened to penetrate a populated site, the probably is still low that any direct evidence that would suggest even a glimmer about what the surface world is like would be retrieved. So from the standpoint of discovering any evidence of life underground, our current exploration is relatively meaningless.

This is not to suggest that if they exist, underground humanoids would have developed civilizations as sophisticated as ours, or have populations of any size approaching our current world population of over five billion. This brings up the question of how many creatures, if they exist, may we be talking about. As already mentioned, water accounts for over 70% of the Earth's surface, ice, another 10% and permafrost yet 15% more, making settlement by humans very sparse indeed. So, the overwhelming portion of us occupy less than 5% of the globe's actual surface area. This is hard to believe, since for most of us, whether we be urban or rural, there seem to be people everywhere.

Geographers will tell us that overpopulation is defined in large part by culture, technology and expectations, and that while there is acknowledged overpopulation in a great many places, this is in reference to the practical availability of usable areas given our current technology and cultural habits. If even less than one tenth of one percent of the Earth's underground potential proves to be habitable, quite sizable populations might exist.

The total land area of the world is 57,800,000 square miles.

The total surface area of the globe, including all of the areas covered by water is roughly 196,950,000 square miles. If one percent of the potential underground proves to be habitable in some way, this is at least 1,900,000 square miles, an area more than half the size of the U.S. or Europe or nearly three times the size of Mexico. The population of Europe exceeds 674 million, the U.S., 238 million and Mexico 78 million.

So we can see that a great percentage of potential underground space is not needed for there to be sufficient room to allow for huge populations. And this assumes average population densities similar to areas that we do not generally consider to be overpopulated (although there are undeniably some cities that are). Even if far less than one percent of the underground were to be populated, it would still be possible for there to be concentrations of humanoids in areas as large as France, Spain or Japan, and even if as little as only one-tenth of one percent is usable, that would still be more than 19,000 square miles - three times the area of the Hawaiian Islands and could house a population of nearly 16 million at the same population density as Japan. What's more, this assumes "single layer" population densities which is generally the rule on the surface, except for high-rise office buildings and apartment complexes in larger cities. In a different culture, layered or multi-story occupancy may be the norm and even greater population densities than those described may be acceptable and functional. This especially would be the case underground, since it is unlikely there are vast areas which measure many hundreds of feet from floor to ceiling.

This is not to suggest that areas or populations anywhere approaching these numbers exist. If at all, the figures are most likely far less. These examples are only to illustrate the potential. Even if the actual numbers of creatures found to dwell beneath the Earth were to be in the hundreds of thousands or tens of thousands, quite advanced civilizations could develop. The size of a population does not have a direct bearing on the level of technological advancement or cultural sophistication. If raw materials and energy sources were available, and sufficient societal cohesiveness exists balanced with an optimal level of diversity, very small populations can become quite refined.

Our history as a people has been a pattern of advance and regress in fits and starts. Although we may feel that the history of our world has been one of great progress, it is by no means uniform in all areas or for all of our peoples. As a whole we may be progressing, but in different places at different times much progress is lost to negative and even destructive change. We gain in some areas and loose in others. The argument can be made that this is a necessary by-product of vital growth and that stability is symptomatic of stagnation, however we have no measure by which to determine if our rate of development as a people has been optimal or something far less.

It may be that under conditions other than those that confront us on this planet's surface, and with a different biological matrix than we are a part of, considerably more progress could have been gained. After all, we have not effectively come to terms with the biological traits and combinations of conditions which have caused periodically disruptive wars, political oppression, bigotry, hatred and criminal behavior. In short, we may not be doing as well as we might have. We cannot assume that development under other conditions and with much smaller numbers would result in an inferior civilization.

The majority of caves which have yielded most of the information about ice-age cave dwellers are in Western Europe, with some greater concentration around central and southern France. These areas have long histories of continuous settlement, and sites here are readily located and explored. There are many other locations which are now inaccessible but may well have been as used by primitive man.

The term land bridge implies a narrow strip of soil which facilitates a precarious passage between larger land areas. The land bridge between Siberia and Alaska was not such a narrow strip however. At the height of the last major period of glaciation, about 18,000 to 20,000 years ago, Asia and North America were connected by an area of land greater in size than Australia. A massive area stretching from the present-day Laptev Sea, north of Siberia, East, through what is now the East Siberia Sea, across the Bering Strait to north of the Alaska Brooks mountain range, South nearly to the Aleutians and West across the Bering Sea, was entirely a land mass - in itself as large as a continent - which was exposed as the level of the seas dropped 400 feet. This area, now known as Beringia, is an extremely likely location for considerable ice-age human activity, and in the edges of the landmasses now below sea level, north and east Siberia and north and west Alaska, we may find evidence of underground occupation and entry into the Earth's crust.

Archeological evidence confirms that prehistoric peoples traveled over this land bridge, and since Neanderthal and Cro-Magnon Homo sapiens used caves in Europe and other continents, we can be almost certain that any openings in the Earth which could provide shelter in this region also would be used. Although this area lies far North, just below the Arctic Circle, it occupied a broad position between the glaciers which covered Northern Europe and Asia and those which covered most of North America. The climate in this zone may well have been more temperate than Southern Europe or Southern U.S. As these glaciers receded, the water released by the melting caused the seas to rise again and flood areas which had become occupied by man. The inundation was slow in human terms, only a couple of feet per century. However, the rate at which land is overrun by rising water depends on the slope of the land, and some areas which were relatively flat experienced much more rapid submersion. Along the western Gulf of Mexico, for example, as much as 200 feet of dry land per year was flooded.

Although the Siberian land bridge may be the largest single landmass exposed during the last ice age, many others existed which were great expanses nonetheless: A portion of the present-day sea floor off almost the entire southeastern coast of South America, stretching from Rio De Janeiro to Tierra Del Fuego, east to the Falkland Islands at the widest, was also exposed.

The Spitsbergen Islands east of Greenland and due north of Sweden were more than twice their present size when their coastlines were expanded to the South and East. In addition to the western Gulf of Mexico, a large portion of the Gulf north of the Yucatan Peninsula, and a sizable portion of the Caribbean Sea east of Honduras and Nicaragua was also above sea level. A huge expanse of the floor of the North Atlantic east and south of Newfoundland was exposed. The eastern coast of China, just off Shanghai, the Yellow and East China Seas, were then an extension of the mainland. Australia and New Guinea were also joined, and the Philippines and Indonesia were connected by a continuous land bridge to the southeast Asian Peninsula. In North America, almost the entire Hudson Bay was above sea level. In fact, most of the currently submerged portions of the continental shelves were, to some degree, above sea level.

Even Antarctica is not beyond consideration since it too was once a part of the original supercontinent, Pangea, which existed as a single, huge landmass more than 200 million years ago. Australia and Antarctica did not split and begin to drift apart until about 45 million years ago. Fossils of plants and animals show that all the large continental landmasses of today were once joined together. If prehistoric man became an underground dweller much earlier than the last ice age, any of today's continents may be a potential site for underground development, for many ice ages have occurred over the last 4.5 billion years. Modern man (as Cro-Magnon) did not appear in Europe until about 32,000 to 33,000 years ago, but there is evidence of early humans in Australia over 40,000 years ago, and more primitive precursors, even earlier than that. By this time, Australia was already separated from the Asian mainland and it is assumed that they must have crossed at least 60 miles of open water to get there. It is even possible that some early human ancestors migrated to, or under Antarctica when the seas were at their lowest.

If the appearance of subterranean humanoids is very much like that described in chapter 6, and that which is common in the many accounts of abductions, then it is likely that their separation from surface humans occurred much earlier than the last ice age. The physiological differences between this description and the appearance of late Neanderthals and Cro-Magnons is probably too great to be accounted for by the scant 20,000 - 30,000 years since their cohabitation of western Europe and the last ice age which followed. However raising and lowering of the sea level occurred with every ice age, perhaps not a much as during the last one, but nearly so. And with each advance of ice, water levels lowered, offshore areas were exposed, only to be covered by water again when the glaciers gradually receded. Any one of these cycles, which took close to 100,000 to complete, would have allowed humans to enter the Earth's crust at points which are now difficult or impossible for us to explore. Long before this, pre-human hominids may have ventured below through sites which can never be located. This would have allowed the time required for full adaptation to the environment and would account for the completely different appearance.

Of course, these descriptions may not be entirely accurate. The physiological traits which were discussed in chapter 6 are those that would seem to result from long adaptation to cave environments. If the time since their imprisonment underground has been as short (in geological and evolutionary terms) as since the last ice age, those traits would not yet be fully developed. But also must be considered is the fact that the perception and memory of startled and frightened people tends to exaggerate unusual aspects of their experiences. In the accounts of close encounters and abductions, even if truthful, only slight differences from usual human appearance would stand out in the memory as gross distortions.

In any event, from a strictly geological standpoint, it is beneath the continental shelves that seem to be the most likely locations of subterranean habitats. The Earth's crust beneath the oceans is generally much thinner than beneath continental masses and is more regular in its composition. Greater irregularity and greater thickness suggest more possibilities for the existence of cavities.

The original passageways into the continental shelves are probably no longer visible through direct observation. Now underwater, silt, plant life, undersea currents and other geological disturbances likely would have caused the concealment of all but the largest openings.

We have only a fragmentary picture of the origin of humans and their dissemination to various continents. Archaeologists and Paleoanthropologists can piece together the course of early human development and migration only through the physical evidence found in scattered sites around the world. Good science must not interpret events far beyond that for which there is evidence, so at this time, the best theories are that early humans, Homo habilis (the immediate precursor of Homo erectus) and Homo erectus probably originated in southeast-central Africa around one to two million years ago. Based on similar fossil remains, the assumption is that migration was to the northwestern African coast and north to the Suez area, around the eastern Mediterranean crescent and westward into Europe.

Fossils of Homo erectus have also been unearthed in northeastern Asia which date from about 750,000 years. The migration from there was eastward across the Beringean plain from northeastern Siberia to northwestern Alaska during the last ice age, 15,000 to 30,000 years ago. The spread then was South and East throughout North America, and finally to South America. At about the same time, Homo erectus was crossing from southeast Asia to Australia by a land bridge similar to that which existed between Siberia and Alaska.

The oldest fossil records then, are in southern and southeastern Africa, with some sites relatively near the coast of the Indian Ocean. If penetration into the Earth was made by these earliest humans, then the continental shelf just off the southeast African coast, near Madagascar, is a prime location for investigation. If later humans - Homo sapiens - were the first to migrate permanently underground, then the area of continental shelf north of Siberia or northwest of Alaska might be the best places to study. The migration from Asia to Australia might also have afforded opportunities for underground migration. However, these are only guesses based on patterns of population movement indicated by archaeological finds to date. Certainly more discoveries will be made which will expand and clarify the current picture.

The development of humans from ape-like hominids to Homo habilis, then to Homo erectus and finally to Homo sapiens, occurred almost simultaneously with the Pleistocene epoch and the many periods of glaciation and fluctuations of sea level. Throughout almost all of human history, during the time when we began migrating all over the world, our development was inextricably bound up with the erratic changes in climate and sea level, and the alterations to the environment that took place. Any areas of landmass which were alternately covered with ice and then exposed, or parts of the continental shelves which were above sea level and then submerged during the last million and a half to two million years are possibilities.

CULTURAL AFTERSHOCKS

What would be our response to discovery and confrontation with humanoid creatures from beneath our feet, our own relatives no less! Our whole orientation and subliminal mental preparation for contact with other species has been for contact with aliens", true aliens - from a world far away. In our imaginations we have envisioned them as different, but still human-looking enough so as not to generate complete revulsion. This visualization is important psychological groundwork for diminishing the stress of new experiences. But this kind of discovery would be absolutely unexpected. And what would our attitudes be about the fact that we are related to such odd-looking individuals? Would we be amused, repelled, extend our hand in friendship or immediately feel the need to reject or dominate. How would we react to the fact that we alone are not the sole heirs to planet Earth and that others may have entirely different ideas about the direction in which technology should progress?

Our tendency as a people (and all cultures are guilty of this) is to relegate minority groups or individuals with whom we must deal on a regular basis, to lower status. Other minority figures such as royalty, entertainment celebrities, politicians and so forth can be accorded positions of distinction since we do not deal with them and so they are personally non-threatening - their spheres of activity do not encroach upon ours. And even if minorities are demonstrably superior in some way, values are reinterpreted in such a way as to cast the minority in a less attractive light or define our association with them in terms that allow us to maintain our self-esteem. Therefore, our view of earthly aliens would probably begin much the same. They would be seen as an oddity, something impersonal which did not directly affect us, and so differences would be no reflection on ourselves. Although this encounter would be, for some, a disturbing experience. Worth noting is the stir that evolution theory caused, and continues to cause in some quarters. Some people may be horrified to be ancestrally associated, however remotely, with such alien creatures.

If the aliens posed no threat to us - and it is unlikely that they would - we would probably, for the most part, be charmed and excited about the discovery. We would naturally be curious and some would want the first opportunity to view their world. Geologists would want to know more about the Earth's interior, anthropologists and archaeologists would want to learn all they could about the evolutionary connection between them and us. Sociologists and psychologists would want to understand their culture and relationships, while various physical scientists and industrial technicians would want to study their science and technology. Military leaders would be concerned with maintaining a secure footing to our relations and politicians would grapple with proper protocol and scramble to put the most politically profitable face on the new turn of events. Businessmen would seek to establish trade and develop new markets for merchandise and services, and of course the news industry would cover the breaking events with the usual array of experts to interpret the significance of it all. The rest of us would watch. In short, every constituency would seek to get whatever benefit they could from it.

This somewhat cynical, hypothetical vision is not all that would happen, of course. Genuine attempts at understanding would be made and some agreements would be established to govern future relations. Wiser heads would immediately recognize that the situation requiring coexistence of two intelligent species would not go away and would have to be dealt with. Our cognitive world would have expanded and we could never go back to the way it was.

The aliens would be allowed to visit and freely travel to most nations, with the proper papers of course, and superficial courtesies would be extended. Humans would also probably be allowed into their world with similar experiences. We humans would marvel at the strangeness of them and their accouterments as, no doubt, they would at us. We would both learn from each other with mostly beneficial results, so long as upper and lower world sovereignty were observed. All would probably be smooth for a while, if neither side felt a challenge to their autonomy.

This "honeymoon" in the new relationship would only begin to crumble as the aliens began to assert themselves and move into relationships that were more than just formally polite. The usual resentments and reprisals would work their way into this new, strained human alliance with unpredictable outcomes. In addition to the strain of interpersonal relationships, inevitably, one side would desire or require something the other had, either access to needed space, minerals or mineral rights, water, materials, technology - whatever, and the conflicts which have marred our history (and perhaps theirs also) would erupt. It is not likely that the disputes would be terminal, however. None of ours has proven to be in the long run.

The aliens would, in time, assume the same status that other religions, cultures, races and nationalities have. They would eventually become a part of the family of man with all of the positive and negative consequences of kinship - family fights and all.

Long-term relations would likely follow much the same pattern as with our other neighbors; generally stable and smooth, interrupted by periodic discontent and perhaps even hostilities. An interesting feature of these unusual circumstances would be the matter of ownership, if that could be the correct term. Of course no one would "own" these people, however the country within whose borders the opening which connected our world to theirs was located would control access. Members of other nations may not be allowed to visit the site, giving "national security" as the justification for denial. For that matter, the host nation would control the aliens' access to the surface world. Just as with underground minerals, the exact source or their origin matters less than the location from which they are accessed. Other openings to their world might be drilled, either by us or by them (ideally, arranged by cooperative agreements, since unilateral action by either side could result in damage to the other's world). If the entryway to the surface happened to be from the water, territorial waters may or may not be a factor, depending on the distance offshore. Conflicts could develop as to the right of either party to deny access to the other's domain.

Legal systems are concerned with describing permissible behavior so as to reconcile individual urges with the need for stability and the common good within the social and physical context, codification of those regulations, punishment for breaches of the regulations and the resolution of conflicts which arise between individuals. Legal text and practice varies widely throughout our history and in different cultures. We can assume that an underground society would face the same issues, but because of the markedly different physical and social context, their legal system would probably seem strange or shocking to us.

In the most basic and personal matters such a sexual activity, marriage, descendence and personal property, our own laws and practices vary enormously. These variations range from single to multiple spouses and from the minimal personal property of communal colonies to the unlimited personal acquisition allowed under capitalism.

An alien culture, concerned with maximal unity and cooperation may allow complete sexual and reproductive freedom, with population growth curbed by non-treatment of the ill, as already described. In such a society, paternity would not be a personal matter and all members of the society might bear a share of the responsibility for offspring. This would eliminate the idea of ownership of children. Infanticide may also be practiced if unlimited births were allowed. On the other hand, the marriage union could be even more tightly controlled than in our society with only one partner allowed for life, with severe punishment, even death, for infidelity. Child-bearing may be permitted only by consent of a governing body and the offspring either allowed to remain with the birth couple or delivered to one or more outside agents for rearing.

Personal property might be nonexistent. Everyone may have access to nearly everything produced by the society. Certainly the environment as we might imagine it would not encourage the freedom and personal acquisition of possessions that we are accustomed to. A home with surrounding land or yard would not exist underground (as it does not now exist for many residents of large cities). Close confinement would have led to laws and a social structure which minimized the problems unique to them.

Would the form of government be democratic, as has been the trend of governmental evolution in our world? This has been one of the major struggles for all people throughout our history. For an underground society however, it may be counterproductive. Autocratic rule may better suit their circumstances. Unity of purpose is greatly simplified when dissenting views have no influence. If all members of the society are subordinate to a single authority, tremendous results can be obtained, as witness the rise as a world power of Germany under Hitler. The danger of such subjugation, of course, is that without the checks and balances of different viewpoints, where none questions authority, eccentric or dysfunctional policies that serve the personal whim of the leader will be pursued instead of a more common good.

Rather than a culture which reflects some expression of the collective values of the members of the society, an underground people might more resemble a cult following, with all blindly following a charismatic leader. A tacit assumption in our belief system is that, in the long run, survival and progress depends on the willing participation and contribution of the best ideas of at least most members of the society, and that radical, autocratic organization is too rigid and limited, and ultimately doomed to fail, either out of excessive ambitions of the leader or from the lack of infusion of new ideas. We have traditionally made a similar assumption in considering the possibility of contact with extraterrestrials; that in order to master the technologies necessary for interplanetary space flight, aggressive impulses and destructive behaviors would have been conquered. We can only hope that this is so.

ASSIMILATION

Would alien creatures, even if they shared the same origin as ourselves, ever be fully assimilated into our society? For it is our society that would absorb them and not the other way around since they would be the minority. Would they ever be accepted as full participants in all aspects of our social life? Would we even view them as human, or at least human-like?

Our disregard for intelligent beings that are much different from ourselves has considerable precedent. We humans seem to have difficulty in identifying with other life forms unless they look, speak, act, behave and even think like ourselves. For not only do we exhibit discriminating and rejecting attitudes towards those with visibly different appearance, we often demand, as a condition association, that we even think a certain way - share the same views and articulate whatever prevailing or politically correct ideas are the common ones required of whatever group to which we seek to belong. For all our talk about open-mindedness, respecting differences, free speech, open expression, and valuing individual freedom, we actually have serious problems accepting that which is different.

People or groups which are different provides that insecure part of ourselves, that part which needs reassurance and bolstering, a target for devaluing. And the devaluing process requires that the object of our derision be enough unlike ourselves so that we are not hurt in the process. The mechanism is twofold: First, in an attempt to elevate ourselves in our self-esteem, we characterize the other party as deficient (this can take the form of contempt or masquerade as benevolent, sympathetic and ultimately condescending understanding). Secondly, we form a reaction which distances ourselves from the target of the devaluation. We try to see ourselves as unlike them as possible, and this psychological model becomes the basis for a reaction formation which constrains our own potential. It is an ultimately destructive process, both for the individual and for the society.

The exception to this reaction (and it is not really an exception at all) is if the minority is in such a subjugated and subordinate position that they represent no emotional threat to the security of the dominant party. It that circumstance, the minority actually is highly-valued and well-treated, since their presence serves to boost the sagging self-esteem of the dominant party. This is really a form of condescending benevolence already mentioned.

This process happens individually but is supported and practiced on a collective basis. And although the better sides of ourselves work to contain these negative impulses, they are present, to some degree or other, in virtually all humans. It is part of our frailty.

In considering the potential for racial prejudice against an underground species, we cannot overlook the possibility that they might consider us quite inferior. We naturally tend to assume that any minority group will be impressed with us and feel inadequate simply because their nation may be smaller, or less technically advanced, or predominately of a different color, or their religion is not Christian, or because they dress differently, or their marriage customs differ - or whatever. This kind of thinking has even dominated the official policy of our country in some cases, as evidenced by the archaic doctrine of Manifest Destiny.

It's true that an arrogant mindset often invites the minority party to participate in a relationship by accepting the "inferior" status. However, this dynamic if fed by the need of people to establish relationships, even at the cost, in some cases, of diminishing one's sense of worth. An alien race may not have similar emotional needs. The need for relationships may not be the overriding factor in their psyches. They may well feel that their culture is far superior and not be in any way intimidated by our machines, open spaces or enormous populations; or by our traditions, principles or politics. They may value cohesion more than diversity: A small, compact, efficient society in which each individual is called upon to participate fully, one in which every contribution is a positive component of a shared sense of purpose may be their goal. They may see us as extravagant and wasteful, not only of resources but of human potential as well. They may not wish to associate with us.

They may find our physical presence offensive. Our bodies may seem to them large, course, and smelly. Our movements may seem gross and clumsy. They may feel that they are refined in their personal actions and coordinated in group behavior. Our appetites and our capacity for indulgence may seem incomprehensible. The thought of physical contact with us may seem repugnant - at least at first. But if centuries of encounters and assimilations of different cultures and races has taught us anything, it is that when individuals relate to each other on a personal level, differences tend to become surmountable. It large part, it is the association and identification with the group that presents the greatest barrier to successful association and assimilation. After individual unions produce offspring which cannot be identified physically with a particular minority group, and as identifiable cultural behaviors are diminished, prejudicial attitudes have no raw material from which to create biases. Ultimately they would be assimilated.

We are a long way from integrating all races and religions which presently exist on the planet now, although we are moving fast in that direction. Perhaps there would be genetic incompatibilities that would prevent mixed parentage. If that is the case, it may be that by the time such events might come about, genetic engineering will have progressed to the point that artificial in vitro fertilization and incubation would overcome whatever difficulties were present, and hybrid humanoids would be possible. If hybrids were created artificially, it might obviate many of the objections some would have against mixed couples.

While it may be socially unacceptable for offspring to result from the union of an alien and a terrestrial, if it were to be possible, it would happen, sooner or later, by natural or assisted means. Consider the examples of surrogate parenting, sperm donors, and artificial and in vitro fertilizations as evidence of the unique partnerships which can be arranged when the outcome suits the needs of both parties. Just a few decades ago, these methods of conception would have been unthinkable for most (and still are for many). So despite any objections or official sanctions that might prohibit it, offspring that are of mixed genes will be born, or created. It will then be the task of both societies to deal with the reality, just as is always the case when technology or conditions expands the possible.

Laws which now cannot seem to keep up with advances in medicine, technology and changes in social behavior, will be strained to the breaking point by new challenges. Entirely new definitions of rights, citizenship, lineage, taxation, inheritance, custody, parental rights and responsibilities, community participation, and so on will have to be revamped to include a new group of individuals. Up to now, our laws have had to deal only with individuals that are indigenous to our world, and that has been hard enough. Governments and legal systems will then have to handle increasingly complex situations brought about by technological advances and also by the assimilation of a truly new group of people.

Home at Last

The argument has been made that the natural course of events in the progress of the world includes the ultimate merging of all races and nationalities. Through interbreeding, children will exhibit fewer identifiable traits of their parental ethnic backgrounds in subsequent generations and the entire world will ultimately become one race. This is often described as not only inevitable, but a positive thing, for valuable traits from different ethnic origins will blend to enhance the human species.

America has often been described as a "melting pot" since virtually all of its people either emigrated from other nations or are the descendants of immigrants. Very few of us are descendants of the indigenous inhabitants. This is usually characterized in a positive way, it being pointed out that diversity of cultural backgrounds and ethnic origins provides our society with unique richness. This philosophy is probably more a result of economic expediency than of an insightful world view of humanity, for the continual influx of individuals who immediately function as an underclass, willing or forced to accept hard work at low wages, is decidedly attractive for policymakers responding to the needs of industry and commerce. Nonetheless, there may be some truth to the idea and it certainly has become a common assumption and a part of our national ethic.

Of course this sort of mixing has been going on for a long time - as long as humans have existed, in fact. One of the theories about the disappearance of the Neanderthals in Europe during the last ice age, was that they were absorbed, through unions with the "immigrant" Cro-Magnons and lost their identifying characteristics through successive generations. This theory is supported by the fact that there seems to be little evidence of mass extinction, either by natural causes or through conflict with the encroaching Cro-Magnons. That may or may not be the case, but there is ample evidence for archaeologists to offer this as a viable hypothesis. Recorded history documents that sexual unions between individuals of different races, either sanctioned or clandestine, have been significant elements in the human biological matrix for millennia.

Whether one accepts this premise of improvement of the species based on the integration of varied ethnicities or not, it is interesting to speculate on the possibility of ultimate assimilation of an underground people by others on our planet. For if they exist, and if genetic mixing is possible, either naturally or artificially, this is the logical direction of events, barring some catastrophic interruption. How long this might take is too difficult to imagine, for complicating the whole matter is the fact that our world population will be curbed by one means or another, long before the natural generational sequence would allow complete assimilation.

At the current rate of world population growth, we simply do not have the possible number of generations remaining to us as has passed since even the last ice age. The Earth is approximately 4.6 billion years old. In geologic and stellar time, that's nothing. In human terms however, it's staggering: 1.6 billion years ago the first oxygen-breathing life appeared on Earth. The dinosaurs roamed 100 million years ago and by 65 million years ago they vanished in a mass extinction, the causes of which are still not understood. Humans as we might recognize them, distinct from more ape-like hominids, appeared one to two million years ago. It took from then until 1850 for world population to reach 1 billion. It has taken only 100 years for that number to quadruple.

Either some horrorific population-limiting events will transpire (viral epidemic, nuclear war or accident, terrorist sabotage, environmental catastrophe, massive crop failures, meteorite strike, take your pick) or population growth will become self-regulated by governmental policy and/or social convention.

We are seeing evidence that when nations become sufficiently developed economically that large families are no longer an advantage as they now are in many third world countries, and population growth tends to decline. Added to this is the fact that the rising social position of women in advanced countries also tends to discourage large families. There may some hope that world population can be contained by civilized means rather than the brutal controls nature will impose.

What actual checks to population growth come into play is anybody's guess. When it stabilizes however, the degree to which subterranean populations will be integrated with all world's peoples may be a matter of natural selection, social control or governmental regulation. Philosophically, it will represent a reunion, symbolically and actual, of long separated members of the same family.

If it is to be our destiny to escape Earth and become inhabitants of the larger universe, then it may turn out that our exploration and eventual migration into space will require some physical or mental traits which we currently do not possess but that subterraneans do. Union with, or at least cooperation with underground humanoids may be part of the penultimate events that will make the undertaking more successful.

Certainly the ability to function in low-Oxygen environments would be an asset on long space voyages. Smaller physical size would also be useful, not only because of the limited space available on spacecraft but would also diminish the effects of G-forces during long periods of sustained acceleration and deceleration. Other faculties such as low-frequency, vibration-sensitive hearing or vision sensitive to electromagnetic bands other than visible light might prove more valuable than our senses when it comes to firsthand exploration of other planets. And of course any species long accustomed to confinement in a cramped environment, a species which had adapted to it through evolution, would more easily endure life aboard a small spacecraft. It may be that, paradoxically, underground people would be better-suited to leaving Earth than we are. Living within the Earth, they have in effect, been riding inside a spacecraft hurtling through space throughout all of their existence.

Perhaps there would be other benefits to associations with underground beings; benefits of a more down-to-earth nature. As pointed out, the underground environment is surely more limited in resources as well as physical space. Therefore efficiency, recycling, minimization of waste, nonpolluting technological processes - all increasingly needed elements in our societies - would be guiding principles in their habits. We could learn a lot from them. We often naively assume that our collective scientific acumen and technological know-how is advanced enough that solutions to contemporary problems are easy to come by, and that the main reasons good ideas are not utilized are political and economic in nature. Although this is frequently the case, we are still limited by failures of imagination and lack of fortuitous accidents.

Throughout human history, whenever one culture first encounters another, there are things which each can learn - benefits to be gained. The unfortunate reality is that often those things are not recognized and taken advantage of. Too often the rush to exploit and the fear of loosing identity obscures cultural practices which meet some human needs and could be usefully incorporated; but that is a failure of the principals involved not of the principle itself. Still, there is always some beneficial fallout from the merging of cultures, and although it frequently brings great difficulty, in the long run the advantages outweigh the losses.

If we accept that biological diversity is a fact of nature and that the strength of sexual reproduction (as opposed to asexual reproduction of some forms of life) is that it offers greater potential for change and adaptation, improving the chances for survival, then assimilation of aliens is a completely natural part of the world's growth. If countries whose populations are rich in diversity benefit from it, then a world which integrates its subterranean and terranean peoples will also be the better for it. There may be mental talents, sensory processes, physical traits and cultural features that can contribute to the evolution of more capable earthlings. Those attributes which they may possess that are not adaptive to life on the surface will disappear as we pass on to them, our own strengths. In time, we may become one people.

EPILOGUE

The ideas presented in this book may seem so fantastic to many readers that they are apt to doubt my sincerity. They may think that this is simply thinly-disguised science fiction in a non-dramatic format. That is not the case. Although to some extent, all theories are fiction until proved, I do expect that forthcoming events will reveal much of this theory as fact.

The range of these speculations resulted from the "what if" kind of questioning that should be a part of any hypothesis. It may be that a subterranean culture existed but has vanished, or that only a few surface humans managed to survive briefly in underground locations, but never developed a civilization. Whatever the case; whether the circumstances described in this book are shown to be actual or become disproved is a matter for time and additional investigation to determine. Certainly the coming decades will see extraordinary discoveries, many of them about the Earth itself. If the discoveries are not those described here, perhaps they will be even more remarkable. The mysteries of life and of natural phenomena are more incredible than any fiction could be. Nature's capacity to invent far exceeds any writer's imagination. I will not be disappointed if future research shows that my hypothesis has been flawed, for it will surely be replaced by discoveries that are even more wondrous.

BIBLIOGRAPHY

Asimov, Isaac. *Asimov's New Guide to Science.* New York: Basic Books, 1984.

Asimov, Isaac. *Exploring the Earth and The Cosmos: The Growth and Future of Human Knowledge.* New York: Crown Publishers, 1982.

Asimov, Isaac. *Extraterrestrial Civilizations.* New York: Crown Publishers, 1979.

Asimov, Isaac. Frontiers: *New Discoveries about Man and His Planet, Outer Space and the Universe.* New York: Truman Talley Books-E.P. Dutton, 1989.

Bailey, Ronald H. and the Editors of Time-Life Books. *Glacier.* Alexandria, VA: Time-Life Books, 1982.

Ballard, Robert D. *Exploring our Living Planet.* Washington, D.C.: National Geographic Society, 1983

Chorlton, Windsor and the Editors of Time-Life Books. *Ice Ages.* Alexandria, VA: Time-Life Books, 1983.

Clarke, Arthur C. *Arthur C. Clarke's Mysterious World.* New York: A & W Publishers, 1980.

Cloud, Preston. *Oasis in Space.* New York: Norton, 1988.

Cone, Joseph. *Fire Under the Sea: The Discovery of the Most Extraordinary Environment on Earth-Volcanic Hot Springs on the Ocean Floor.* New York: William Morrow and Company, 1991.

Corliss, John B., John A. Baross, and Sarah E. Hoffman. *Submarine Hydrothermal Systems: A Probable Site for the Origins of Life.* Corvallis, Oregon: Oregon State University, 1980.

Erickson, Jon. *Craters, Caverns and Canyons: Delving Beneath the Earth's Surface.* New York: Facts on File, 1993.

Erickson, Jon. *Ice Ages, Past Present and Future.* Blue Ridge Summit, PA: Tab Books, 1990.

Gribbin, John and Mary. *Children of the Ice; Climate and Human Origins.* Oxford: Basil Blackwell, 1990.

Halliday, William R. *Depths of the Earth.* New York: Harper and Row, 1966.

Hambrey, Michael and Jurg Alean. *Glaciers.* Cambridge: Cambridge University Press, 1992.

Hamilton-Patterson, James. *The Great Deep.* New York: Random House, 1992.

Harris, Stephen L. *Agents of Chaos.* Missoula, MT: Mountain Press, 1990.

Idyll, C. P. *Abyss: The Deep Sea and the Creatures That Live in It.* New York: Thomas Y. Crowell Company, 1971.

Kenneth, James P. *Marine Geology.* Englewood Cliffs, NJ : Prentice Hall, 1982.

Kurten, Bjorn. *The Ice Ages.* New York: G.P. Putnam's Sons, 1972.

LeGrand, H.E. *Drifting Continents and Shifting Theories.* Cambridge: Cambridge University Press, 1988.

Lewin, Roger. *Human Evolution.* Oxford: Basil Blackwell, 1984.

Lewin, Roger. *In The Age of Mankind.* Washington, DC: Smithsonian Books, 1988.

Lewontin, Richard. *Human Diversity.* New York: Scientific American Books, 1982.

Lovecock, Jim. *The Ages of Gaia.* Oxford: Oxford University Press, 1988.

McDonald, Thomas R. *The Search for Extraterrestrial Intelligence: Listening For Life in the Cosmos.* New York: John Wiley & Sons, 1987.

Middleton, John and Tony Waltham. *The Underground Atlas: A Gazetteer of the World's Cave Regions.* New York: St. Martin's Press, 1986.

Mohr, Charles E. and Thomas Poulson. *The Life of the Cave.* New York: McGraw-Hill, 1966.

Moores, Eldridge, Editor. *Shaping the Earth: Tectonics of Continents and Oceans: Readings from Scientific American Magazine.* New York: W.H. Freeman, 1990.

Motz, Lloyd, Editor. *The Rediscovery of the Earth.* New York: Van Nostrand Reinhold, 1975.

National Geographic Society. *Mysteries of the Ancient World.* Washington D.C.: National Geographic Society, 1979.

Press, Frank, and Raymond Siever. *Earth.* 3rd. ed. San Francisco: Freeman, 1982.

Skinner, Brian J. and Stephen C. Porter. *Physical Geology.* New York: John Wiley & Sons., 1987.

Smith, Peter R., Editor. *The Earth.* New York: MacMillan, 1986.

Stanley, Steven M. *Extinction.* New York: Scientific American Books, 1987.

Stini, William A., Editor. *Physiological and Morphological Adaptation and Evolution.* The Hague: Mouton Publishers, 1979.

Tattersall, Ian. *The Human, Odyssey: Four Million Years of Human Evolution.* New York: Prentice Hall, 1993.

Waltham, Tony. *Caves.* London: Crown Publishers, 1974.

White, Frank. *The SETI Factor.* New York: Walker & Company, 1990.

INDEX

D

E

F

G

H

I

J

K

L

M

N